...le Life ...ructure

...Research Establishment

...Transport Research Laboratory

Shillpa Singh

Andy Green Faithful & Gould

Andrew Crudgington Institution of Civil Engineers

Dr Das Mootanah CIRIA

BRE is committed to providing impartial and authoritative information on all aspects of the built environment for clients, designers, contractors, engineers, manufacturers, occupants, etc. We make every effort to ensure the accuracy and quality of information and guidance when it is first published. However, we can take no responsibility for the subsequent use of this information, nor for any errors or omissions it may contain.

BRE is the UK's leading centre of expertise on building and construction, and the prevention and control of fire. Contact BRE for information about its services, or for technical advice, at:
BRE, Garston, Watford WD25 9XX
Tel: 01923 664000
Fax: 01923 664098
email: enquiries@bre.co.uk
Website: www.bre.co.uk

Details of BRE publications are available from:
Website: www.brebookshop.com
or
IHS Rapidoc (BRE Bookshop)
Willoughby Road
Bracknell RG12 8DW
Tel: 01344 404407
Fax: 01344 714440
email: brebookshop@ihsrapidoc.com

Published by BRE Bookshop

Requests to copy any part of this publication should be made to:
BRE Bookshop
Building Research Establishment
Bucknalls Lane
Watford WD25 9XX
Tel: 01923 664761
Fax: 01923 662477
email: brebookshop@emap.com

BR 476

First published 2005
ISBN 1 86081 737 8

Contents

This guide is the result of a collaborative effort and was produced by a working group, comprising the Building Research Establishment (BRE), Transport Research Laboratory (TRL), Construction Industry Research and Information Association (CIRIA), Faithful & Gould and the Institution of Civil Engineers (ICE).

The working group was developed as part of a research project entitled: Unlocking Whole-Life Value in Infrastructure and Buildings, developed under the Partners in Innovation programme and funded by the Department of Trade and Industry (DTI).

The working group would also like to express its appreciation for contributions made by Dr Josephine Prior (BRE), Tim Warren (ex-chairman of the steering group) and Christopher Harris (ex-project manager, CIRIA) in the early stages of the research.

The project was led by CIRIA and benefited from the advice and guidance of a steering group of industry representatives.

Working group members

Kathryn Bourke	BRE
Dr Vijay Ramdas	TRL
Shillpa Singh	TRL
Andy Green	Faithful & Gould
Andrew Crudgington	ICE
Dr Das Mootanah	CIRIA

Steering group members

Andrew Crudgington (chairman)	ICE
David Crosthwaite	Davis Langdon LLP
Les Hawker	Highways Agency
Roger Petherbridge	Davis Langdon LLP
Simon Rawlinson	Davis Langdon LLP
Jonathan Simm	HR Wallingford Ltd
Kristian Steele	BRE

People and organisations consulted

The following people were also consulted through workshops and discussions throughout the project and provided valuable suggestions to the working group in developing and improving the contents of the guide:

Javad Akhtar	Hyder Consulting
Brian Bell	Network Rail
Adrian Cashman	University of Sheffield
Mike Clift	BRE
Chris Durrant	Great Yarmouth Borough Council
Jobran Hammoud	Cyril Sweett Ltd
Les Hawker	Highways Agency
Stuart Hercus	Hyder Consulting
Andy Horsley	Ecofys UK
Chris Hounsell	Waterman Burrow Crocker
Andrew Hugill	4ps
John Ioannou	Office of Government Commerce
Paul Jones	Corus Colours
George Martin	BRE
David Meggitt	Holistic Creations Network
Gary Moss	Cyril Sweett Limited
Dr Crina Oltean-Dumbrava	University of Bradford
Stephen Penfold	Jenkins Potter
Roger Petherbridge	Davis Langdon LLP
Bill Prince	Dean and Dyball
Rita Singh	Construction Products Association
Prof Tony Swain	University College London
Huw Thomas	Warwickshire County Council
Dr Derek Thomson	University of Loughborough
Chris Williamson	Weston Williamson

Funders

This project was funded by:

- DTI
- ICE
- BRE
- TRL
- CIRIA
- Waterman Burrow Crocker

Terms and definitions

General terms

Best value is about achieving the nearest possible match to your functional requirements for the best price. It is not about taking the lowest priced option. It is the optimum combination of whole-life cost and quality (or fitness for purpose) to meet the user's requirement.

Clients include major repeat purchasers, small and occasional clients and property developers in both public and private sectors. Clients include both commissioning and non-commissioning clients.

Commissioning clients are those people in the client organisation who are most directly involved in commissioning the construction project. They are in regular contact with the project and the suppliers at every stage of its progress, and they pay for the work. Generally this guide applies to actions by commiss-ioning clients.

Construction means all constructed facilities, buildings and infrastructure. A construction is defined as a physical setting used to serve a specific purpose. A construction may be within a building, or a whole building, or a building with its site and surrounding environment; or it may be a constructed facility, which is not a building, such as a bridge, road or railway for instance. The term encompasses both the physical object and its use.

Functionality The functionality of any construction is the degree to which it enables a business objective to be achieved. Functionality can be expressed in terms of the amount of waste-free value, which the construction adds to a business process.

Key Performance Indicators (KPIs) help a business define and measure progress toward its goals. KPIs are quant-ifiable measurements of the improvement in performing an activity that is critical to the success of a business.

Needs are functions essential to the delivery of business value.

Life-cycle is the life of a project/product/system for its conception through to end of life, decommissioning or disposal.

Optimism Bias (OB) takes account of a systematic tendency by project appraisers to be over optimistic when estimating benefits and tend to understate timings and costs, both capital and operational.

Option evaluation is the process of considering different options in relation to how they perform against chosen criteria for evaluation.

Public Private Partnerships (PPPs) are relationships formed between the private sector and public bodies for introducing private sector resources and/or expertise in order to deliver public sector assets and services. This involves a long-term contractual arrangement lasting typically for 25 to 30 years between a public body and a private sector provider, where resources and risk are shared. PPPs can include different working arrangements from loose, informal and strategic partnerships to design-build-finance-operate (DBFO) type service contracts and formal joint venture companies. PPP is primarily based on the Private Finance Initiative (PFI), which was introduced in 1992 in the UK.

Requirements are functions that are either specific needs or which are desired in order to make the construction project more appealing.

Risk is the likelihood of a specific outcome, at some time in the future, combined with consequences that will follow in a particular context. Usually the outcome is defined as an unwanted event, but it could be an unintended benefit, which is an opportunity to add value.

Risk assessment is a technique for identifying targets of importance to a business and the construction, which enables an estimation to be made of both likelihood and consequences of failure to meet them.

Risk Management (RM) is the systematic process of identifying and managing risks and opportunities for a project or business.

Stakeholders are all members of society who have an interest in construction. We are all stakeholders, and there are many ways in which we experience our stakeholder status, for instance as:

- Residents in our homes.
- Employees in offices, factories or shops.
- Travellers on the roads, railways and in the air.
- By-standers affected by activities around us.

Suppliers are construction providers and they include designers, consultants, building contractors and manufacturers of components.

Value is the final output value from a business or personal activity, which results from a business process, such as manufacturing a product or delivering a service, or a personal activity such as running a home. It has three essential characteristics:

- Value can only be defined by the ultimate customer or user.
- Value is created by the provider.
- From the customer's point of view, providers exist in order to create value.

Values are principles or standards of behaviour ie the things to which we give value and which determine how we behave. Some important personal values are truth, honesty, trust, respect for others and for the environment, fairness, making what we do enjoyable for others with whom we interact, openness, competence, sustainability, balance, harmony, reasonableness. Wherever possible KPIs should reflect the values within a process of all stakeholders.

Whole Life Value (WLV) of an asset represents the optimum balance of stakeholders' aspirations, needs and requirements, and whole life costs.

Life Cycle Assessment terms

Life Cycle Assessment (LCA) is a systematic set of procedures for compiling and examining the inputs and outputs of materials and energy and the associated environmental impacts directly attributable to the functioning of a product or service system throughout its life-cycle.

Goal definition and scoping (BS EN ISO 14041) concerns determining how results will be used, the reasons for carrying out a study and to whom the results will be communicated.

Inventory analysis (BS EN ISO 14041) involves the collection of data about the products or services being investigated, and quantification of relevant material and energy inputs and outputs.

Impact assessment (BS EN ISO 14042) generally involves evaluating data collected on the material and energy inputs in terms of their potential environmental impact.

Interpretation (BS EN ISO 14043) combines the findings from the inventory analysis and impact assessments to allow conclusions to be drawn and recommendations to be made.

Whole Life Costing terms

Whole Life Costing (WLC) is "an economic assessment considering all agreed projected significant and relevant cost flows over a period of analysis expressed in monetary value. The projected costs are those needed to achieve defined levels of performance, including reliability, safety and availability". (ISO DIS 15686-5)

Whole life costs are the costs of all the items/activities that need to be considered in a WLC exercise.

Net Present Value (NPV) is the sum of the discounted benefit of an option less the sum of the discounted costs. It represents therefore a single figure, which takes account of all relevant future incomes and expenditures for that option over the period of analysis.

Multi-Criteria Analysis terms

Multi-Criteria Analysis (MCA) is evaluation by establishing preferences between options by reference to an explicit set of objectives that the decision making body has identified, and for which it has established measurable criteria to assess the extent to which the objectives have been achieved.

Value Management and Value Engineering terms

Function analysis is one of the fundamental techniques involved in a Value Management (VM) or Value Engineering (VE) study. Its purpose is to develop a systematic breakdown of functional requirement, concentrating on the actual needs, aspirations and wants of the client and project stakeholders.

Value Management (VM) is a structured approach to defining what value means to a client in meeting a perceived need by establishing a clear consensus about the project objectives and how they can be achieved. (CIRIA SP129). BS EN 12973: 2000 also defines VM as a "style of management, particularly dedicated to motivating people, developing skills and promoting synergies and innovation, with the aim of maximising the overall performance of an organisation". Applied at the corporate level, VM relies on a value-based organisational culture taking into account value for both stakeholders and customers. At the operational level (project oriented activities) it implies, in addition, the use of appropriate methods and tools.

Value Engineering (VE) is incorporated into VM as a systematic approach to delivering the required functions at lowest cost without detriment to quality, performance and reliability. (CIRIA SP129). BS EN 12973: 2000 also defines VE as the term sometimes used for the application of value analysis to a new product which is being developed.

Purpose and scope

Purpose

This guide to Whole Life Value (WLV) introduces the concept of a) making decisions based on broader criteria than just initial capital costs, and b) taking account of the needs of a broader range of stakeholders than only those traditionally involved in the immediate decision-making process.

The main purpose of the guide is:

- to encourage the move towards using these broader criteria as described above, eg long-term cost impacts as well as other aspects linked to social and environmental impacts).
- for investment appraisal.
- to bring about not only business benefits but also benefits to a wider range of stakeholders.

The guide provides an introduction to the principles of WLV and describes the processes involved in making investment appraisals that encourage WLV solutions for project stakeholders. It considers key questions such as:

- What is WLV?
- Why think WLV?
- Who are the stakeholders?
- When should WLV be considered?
- How can WLV be achieved?

Scope

The guide describes WLV principles for the procurement of buildings and infrastructure assets, eg schools, offices, housing, highways, railways, flood defences, and repair and maintenance schemes. It includes information on some of the existing methods, techniques and tools to support the objective of achieving WLV when commissioning or maintaining an infrastructure or building asset.

The guide also presents a set of case studies illustrating the application of WLV principles on a selection of projects. Techniques covered include Whole Life Costing (WLC), Life Cycle Assessment (LCA), Multi-Criteria Analysis (MCA), Value Management (VM) and Risk Management (RM) processes, based on knowledge and current practice.

However, a detailed account of all the tools and techniques that may be perceived to be embraced under the broad umbrella of WLV and sustainability is outside the scope of this guide.

Target audience

The guide will be of interest to all those involved in commissioning, funding, developing, designing, constructing, operating, using and maintaining infrastructure and building assets. The target audience includes a wide group of stakeholders, all having an interest in the outcomes and benefits created by these assets:

- Clients, owners and occupiers of buildings
- Policy-makers interested in best-value procurement, procurement advisers/professionals/managers
- Funders, financiers, bankers, insurers, quantity surveyors, specifiers, designers and other consultants
- Construction project managers, contractors (prime and specialist contractors), supply team procurers
- Asset managers and asset maintenance professionals
- Users' groups, relevant interest groups and third parties.

For maximum WLV, all those parties interested in the asset need to be involved as early as possible in the decision-making process and project development.

The guide is expected to give stakeholders the confidence:

- to engage with the WLV concept.
- to select and use tools appropriate to their context and circumstances.
- to challenge the barriers on how WLV is achieved in infrastructure and buildings.

How to use the guide

The guide should be used as part of an organisation's decision-making framework for project investment, development strategies and options, and implementation processes. It should be used at different project stages – from the outset of procurement – to identify the issues, principles and tools available to assist in considering WLV.

- The first six sections explain the basics of WLV.
- Sections 7 and 8 provide hands-on guidance in the form of techniques and methods, tools and resources.
- Section 9 identifies future challenges for the application of WLV principles.
- Section 10 provides case studies to illustrate the use of WLV principles and tools.
- The annexes include key WLV drivers in the public sector and a list of tools for reference.

section 1
Value, not just costs

summary
Major procurement in the public and private sector is increasingly being undertaken on the basis of not just lowest capital, or even whole life, costs but 'value'

There are various drivers for this shift from cost to value. Some are mandatory (mainly public sector and affecting their suppliers) and some have come about as clients increasingly recognise that greater benefits can be achieved from the construction procurement process. A summary of the key WLV drivers in the public sector is included in Annex 1.

Value for money drivers in public procurement

A series of initiatives and policy reviews have changed the approach of the public sector to procuring construction projects. These include:

- *Review of Civil Procurement in Central Government* (Gershon Review, 1999).
- *Achieving Excellence in Construction* (Office of Government Commerce [OGC], 1999).
- *Her Majesty's Treasury Guidance Note 7 on Whole Life Costs* (HM Treasury, 2000).
- *Building a Better Quality of Life* (Department of Environment, Transport & the Regions [DETR], 2000).
- *The OGC Gateway™ Process* (OGC 2001, www.ogc.gov.uk and also see Annex 1).
- *Best Value in Local Government* (See Annex 1 for further details).
- *National Audit Office – Modernising Construction*, 2001.
- *Releasing Resources to the Front Line – Independent Review of Public Sector Efficiency* – Sir Peter Gershon CBE, July 2004 (See Annex 1 for further details).

These initiatives directly affect the private sector as partners and suppliers to the public sector.

Local authority practice – the reality

Research commissioned by the Local Government Task Force (LGTF) (CPN, 2004), on counties, unitary authorities, Metropolitan authorities and London boroughs, produced the following insights:

- Nearly a third of those surveyed reported that their authority had let two or three of their last three contracts on price alone.
- Authorities with larger budgets (over £10m) performed no better than those with smaller budgets, in respect of the award of contracts based on quality, rather than purely cost considerations.
- There were some variations in performance by sector; the highways sector performed better than the Buildings sector.
- 70% of the highways sector had let none of its latest contracts based purely on lowest price; 28% had let one or two on this basis.
- 54% of the buildings sector had let none of its latest contracts based purely on lowest price; 42% had let one or two on this basis.

Value for money drivers in Private Finance Initiative and Public Private Partnerships

Since 1997, the number of Private Finance Initiative (PFI) and Public Private Partnerships (PPP) projects in the UK has increased steadily to over 600 projects (International Financial Services, 2003) in industry sectors such as Education, Prisons, Defence, Health, Water and Transport. There is increased focus on service delivery over the long-term (typically 25 to 30 years) to a defined standard and not just delivery of the asset/project. As such, there is a need for an approach that provides value for money in the long term (Figure 1).

PPPs are expected to deliver greater value for money than traditional procurement methods, due to expected efficiency gains and reduction in costs resulting from the sharing of knowledge and skills in design, construction and operation. Risk is transferred to the private sector according to the principle of risk to be transferred to the party best able to manage it. The private sector contractor is expected to provide innovative methods of delivering the service, thereby reducing whole life costs. Improvement in efficiency is expected to be achieved by harnessing the private sector's managerial practices, experience, and ability to innovate and to take risks. A clear understanding of risks within a PPP project is therefore essential for demonstrating value for money. More information on PPPs is available from sources such as: www.4ps.co.uk and www.des.gov.uk/ppppfi.

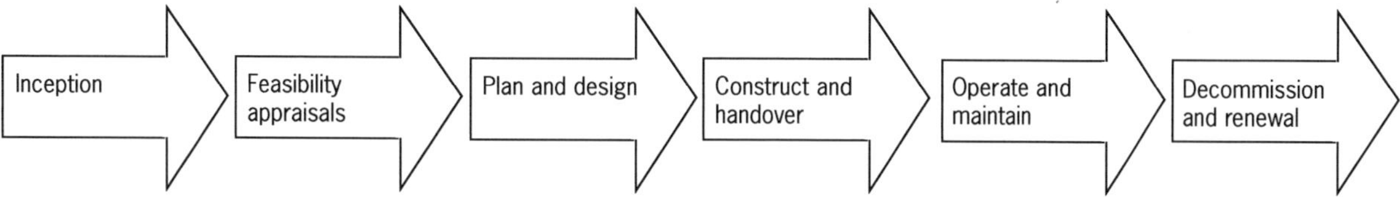

Figure 1 WLV throughout

What about the private sector?

Private sector owners and clients are also increasingly recognising the need to measure projects on value, not just cost. For example:

- *Constructing the Team* (Latham, 1994)
  "Clients should seek to evaluate all tenders on the basis of quality, likely cost-in-use, out-turn price and known past performance as well as price."
- *Rethinking Construction* (Egan, 1998)
  "Increasingly clients are taking the view that construction should be designed and costed as a total package, including costs in use and final decommissioning."
- *Constructing Improvement – the client's pact with the industry* (Construction Clients' Forum, 2000)
  "CCF members want whole life costs to be appraised and the supply chain to commit themselves to build on time, to budget and quality and provide genuine value for money throughout the life of the construction."
- *Accelerating Change* (Strategic Forum, 2002)
  "...advice (to clients) should cover a range of procurement and management options, including environmental performance, operating and whole life costs".

So what?

Although there is a growing recognition for both the public and private sector to consider a whole life and sustainable approach to construction, some clients are still reluctant to do so. To better inform investment decisions, there is a real need to advance the application of tools and techniques that enable the achievement of WLV, in both public and private sector procurement, by:

- exploring alternative opportunities while balancing stakeholders' aspirations, needs, requirements and whole life costs.
- developing best WLV options that will satisfy project stakeholders' objectives and functional requirements.
- optimising the chosen solution, which strikes the most cost-effective balance between initial capital investment, operational and replacement/disposal costs over the required life-cycle of the infrastructure and buildings.

WLV therefore extends the scope of the appraisal process by focusing on more than just the economic aspects of the costs associated with commissioning and operation of an asset and helps the decision makers to consider all the competing factors that drive value.

This guide considers the evaluation of WLV as an iterative 'cradle to grave' process that can be applied to making major investment decisions at any stage in the life of an infrastructure or buildings asset. This may sometimes require the earlier stages to be reappraised, eg feasibility appraisals may require review following cost or need changes later in the project.

section 2
What is WLV?

summary
The WLV of an asset represents the optimum balance of stakeholders' aspirations, needs and requirements, and the costs over the life of the asset

WLV encompasses economic, social and environmental aspects associated with the design, construction, operation, decommissioning, and where appropriate, the re-use of the asset or its constituent materials at the end of its useful life.

WLV takes account of the costs and benefits associated with the different stages of the whole life of the asset. The application of WLV principles is much more than just WLC or Life Cycle Assessments (LCAs), which are integral to the process. The application of WLV includes the consideration of the perceived costs and benefits of some or all of the stakeholders' relevant value drivers.

Stakeholder values, aspirations, objectives, needs and requirements are what drive the search for WLV achievement. This guide will refer to them as the value drivers (some of these values are illustrated in Figure 2).

Stakeholder values differ between different industry sectors and different stakeholders and in practice it is rare for each of these values to be given the same level of importance.

There is no single right answer. Value drivers will therefore need to be identified and prioritised to suit the particular requirements of the specific project or sector. Conflicts are likely between the interests of the stakeholders and there may also be constraints on what can be achieved.

The decision-making process therefore requires clear thinking about the objectives and service requirements of those impacted by the decision over the whole life of the

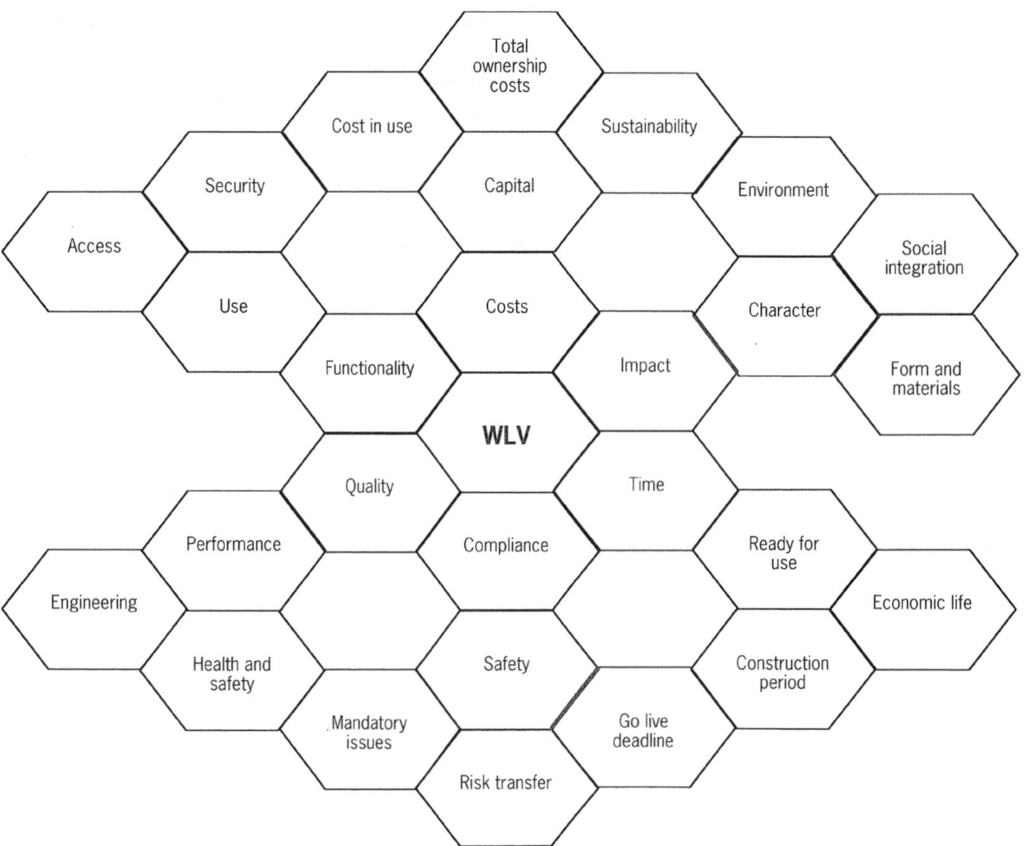

Figure 2 Best Value Framework (European Construction Institute WLV Task Force, 2004)

Figure 3 WLV in our everyday lives

asset and the development of appropriate measures of value.

WLV principles in simple terms

Making any major investment decision is like buying a house. Prospective purchasers are not only interested in the cost of acquisition. They want to know how much it is likely to cost to run and also its likely resale value. Purchasers take into account the cost of finance and income and they want to know if they can afford it.

What they are actually doing is a WLC of different options. When they compare the pros and cons of one house versus another, they will ultimately make the investment decision based on WLV. People know how to do WLV – it is not a complicated process (Figure 3).

So, if we look at what we actually do in our everyday lives, WLV principles are the following:

- Scheduling our requirements – in broad brush terms, defining what we are looking for in order to shortlist options, or filter out, by a simple yes/no.
- For each of the short listed options, thinking value, eg function, build quality.
- For the preferred option, doing a more detailed evaluation before making a decision.

section 3
Why think WLV?

summary

We need to think WLV to take account of competing stakeholder needs to deliver best value solutions

Thinking WLV enables decision makers to identify the investment option that best meets the requirements (often conflicting) of a range of stakeholders effectively and efficiently.

Key benefits of a WLV approach

The key benefits of adopting a WLV approach are:

- involving stakeholders: Helps gain multi-stakeholder support for the selected option, whilst identifying and balancing requirements (needs) against preferences (wants), and future issues, and also identifying and resolving differences in aspirations.
- setting WLV principles: Ensures design development and whole life planning, in terms of specific functional, operational or environmental commitments and use of forms and materials etc, whilst striving to optimise and encourage innovation during the process.
- creating sustainable development: Ensures the appropriate standards, or targets, for sustainable development are considered.
- influencing the future: Uses WLV at the inception and feasibility design stages, to ensure that option evaluation is robust and also economically viable from a whole life-cycle perspective.
- realising best value: Implements effective and efficient use of all resources and tools available throughout a project's life-cycle.

WLV thinking requires a framework for structuring and co-ordinating the decisions and choices to be made when investing for the long term. Such a framework:

- identifies competing value drivers and develops value measures for prioritisation purposes.
- requires the use of accepted WLC and risk and VM tools and techniques, at each stage of the project life-cycle.
- provides the mechanism for evaluating alternative investment options at all the key decision-making stages of the procurement process, including whether the project is needed.
- supports benchmarking between comparable practices to achieve continuous improvement and innovation (WLV encourages benchmarking because all the costs and benefits are being measured, the basis of benchmarking).
- identifies opportunities to agree a basis to trade-off between the value drivers and align stakeholders to make effective and efficient decisions
- identifies the best value solution where there are alternative means of achieving project stakeholders' particular objectives.

Put another way, the ultimate success and future implications of any investment decision will depend on:

- how well the decision makers take account of not just construction costs, but the total life-cycle costs.
- how well the decision makers address stakeholder requirements, eg time, quality, environmental and sustainability issues.

section 4
Who are the stakeholders?

summary
Stakeholders are individuals or groups of individuals who influence and/or are influenced by the outcomes of the project

The range of stakeholders will be different for different types of projects and can include clients, developers, owners, funders, occupiers, managers, contractors, designers, the supply chain, users, neighbours and the general public (Figure 4).

Stakeholders' competing value drivers

Multiple stakeholders will have:

- different and often conflicting values, interests, objectives and requirements.
- different levels of involvement at various stages.
- varying levels of influence on the outcome.
- competing value drivers.

Table 1 provides an example of stakeholder influences on achieving WLV throughout the life-cycle of a project. It can be used as a framework to assess the stakeholders' degree of influence at different stages of their own infrastructure and building projects. Different projects will have different stakeholders and a different level of influence and involvement.

How can value drivers be aligned?

The use of a VM process can assist in aligning stakeholders' objectives and/or value drivers. The challenge in achieving WLV lies in defining measures of value appropriate to the particular project that address the requirements of the diverse stakeholders.

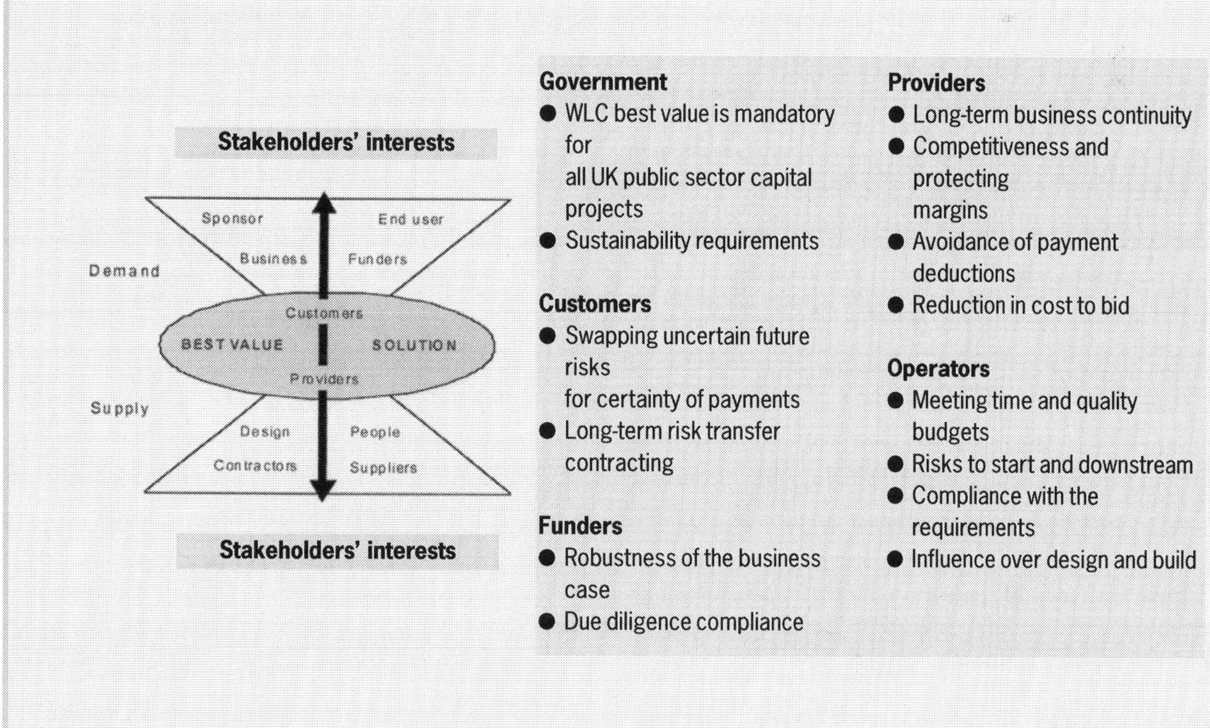

Figure 4 Competing value drivers

Table 1 Stakeholders' indicative involvement and influence in a project's life-cycle

	Inception	Feasibility appraisals	Plan and design	Construct and handover	Operate and maintain	Decommission and renewal
Public and private sector commissioning clients and developers	***	***	**	**	*	**
Users' groups	***	**	*	*	***	*
Occupiers	**	*	*	*	***	*
Policy-makers and planners	***	**	*	*	*	**
Procurement advisers and managers	***	***	***	**	**	*
Funders, financiers and bankers	**	**	*	*	*	**
Insurers	**	*	*	*	**	**
Quantity surveyors	***	***	**	**	*	**
Specifiers, designers and engineers	**	***	***	***	**	**
Construction project managers	**	**	**	***	**	***
Contractors (main and specialists)	**	**	**	***	**	***
Materials suppliers	**	*	**	***	***	*
Asset managers and asset maintenance professionals	***	**	**	**	***	**

Stakeholders' level of influence and involvement for achieving WLV: *** high influence and involvement; ** medium influence and involvement; * low influence and involvement

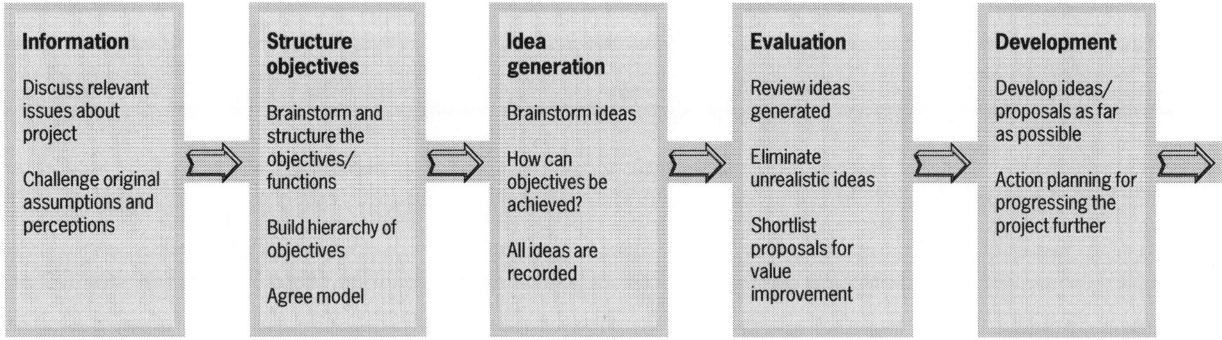

Figure 5 A VM process (adapted from CIRIA SP 129, 1996)

A typical VM process is illustrated briefly in Figure 5. Figure 5 illustrates the activities that stakeholders would be involved in at successive stages of the process, which is implemented through multi-stakeholder workshops. More details on VM and RM processes are discussed in Section 7 – Techniques and methods.

By developing and structuring objectives and functions, stakeholders gain a better understanding of what the project is aiming to achieve. This enables clear and unambiguous definition of the functionality requirements for the asset in accordance with business and stakeholder needs.

For example, aiming for high level outcomes to: reduce energy costs.

- reduce capital costs and operating costs.
- reduce future risk.
- create a healthier and productive environment.

These outcomes can be used to focus priorities of different stakeholders towards using the same strategy.

section 5
When should WLV be considered?

summary

Think value from the beginning and through all stages of the project life-cycle

Value is relevant to the whole life-cycle of project development and should be considered at each decision-making stage. For any project, whether commissioning a new asset or major maintenance/refurbishment of an existing asset, the best time to commit to optimising WLV is the earliest stage when the business case for the investment is being examined. (Note that different procurement routes have different terms for the strategic business case stage, and the options appraisal early stages – public sector Gateway™ terms may be more familiar for some readers.)

WLV throughout the project life-cycle

The impact of timing on value and cost illustrated in Figure 6 shows that the potential for value improvement and cost reduction is highest at the inception stage when there is greater scope to ensure that project objectives are aligned with the interests of all stakeholders.

The early stages of the project will normally allow a high level definition of the project, a broad estimate of the costs and an understanding of uncertainties. As the project develops, and design details are refined, more reliable estimates of the initial and future costs of alternative options become possible.

At each stage, the evaluation can be refined to generate options more closely aligned to user needs and estimate their whole life cost and benefits more accurately. This leads to the identification of the best value for money option, ie one that best meets users' functional requirements at a lower whole life cost.

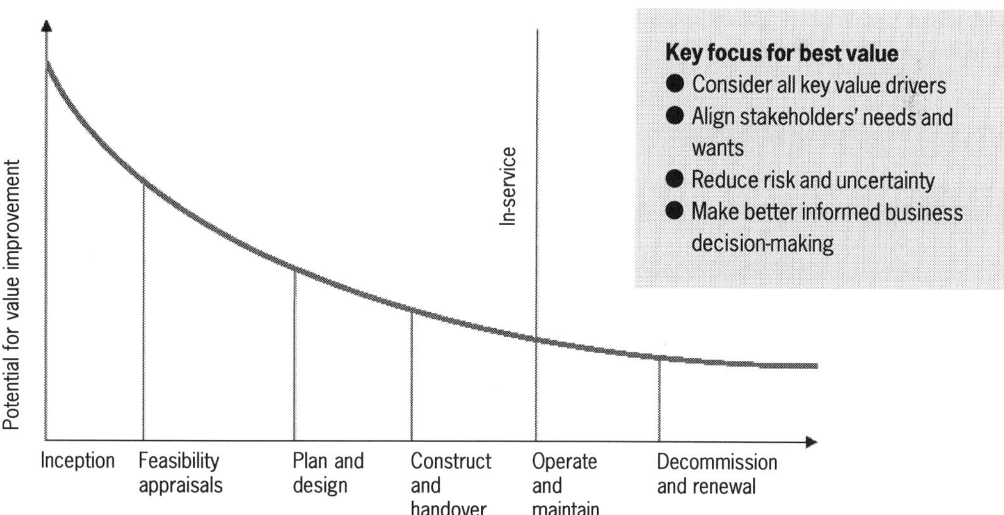

Key focus for best value

- Consider all key value drivers
- Align stakeholders' needs and wants
- Reduce risk and uncertainty
- Make better informed business decision-making

Figure 6 WLV throughout the project life-cycle

section 6
How is WLV achieved?

summary
WLV is achieved by systematically assessing project options against stakeholder requirements

Key pillars of WLV evaluation

The key pillars of the evaluation process are:

- identifying stakeholders, their needs and their relationship with the project (eg provider, user).
- defining the objectives in terms of the outcome required by the stakeholders.
- developing alternative (and innovative) solutions to meet the objectives.
- identifying the best value option, that is: optimum alignment of whole life costs and ability to meet defined outcome requirements.

Box 1 illustrates the evaluation process involved.

Generating options and assessing their value relative to outcome specifications is at the heart of WLV evaluation and should be used at all the different decision-making stages of the procurement process. This could include the use of techniques such as WLC, LCA, MCA, VM and RM.

The overall objectives of the project need to be considered or revisited at each stage of the process to ensure that they will be achieved efficiently and effectively.

In practice the best value solutions will require a compromise to be reached on various issues.

For example:

- Budget constraints.
- Availability of funds.
- Value conflicts between stakeholders (different stakeholders' aspirations, needs and wants).
- Policy restrictions.

Decision support mechanisms can rank options in order of their relative values in respect of defined criteria. For example, pre-set budget targets, minimum acceptable standard and return on capital. This can enable trade-offs between conflicting values, eg higher capital costs but reduced disruption to users, and can help to choose the best value option.

Box 1 Evaluation process

Stakeholders

- Who are the stakeholders?
- What are the requirements and roles of the different categories of stakeholders?
- What are the degree of influence and involvement? (See Table 1 in Section 4.) Early stakeholder involvement is essential.

Functionality and performance

- What is the business need?
- Is there a need to build/construct at all (do nothing option)?
- What is the building or constructed asset for?
- What tasks does it need to facilitate?
- What is the outcome required?

Performance and time

- How does the facility or asset need to behave in use?
- How long does it need to last (service life)?
- How optimistic are the assumptions made? (See optimism bias [OB] guidance on page 23 under Risk Management and Value Management processes)

Risk

- What are the risks?
- Who will manage the risk?
- What are the costs involved?

Environmental sustainability

- What impact will the facility/asset or project have on the environment and sustainability issues?
- Can adverse impacts be reduced?
- Can the impact be costed? What is the cost/value of the impact?

End of life issues

- How can the facility be decommissioned, re-used or recycled?

Cost

- How much will it cost to build?
- How much to operate and maintain?
- How much to renew/dispose?

A step-by-step process

The development of WLV solutions is a multi-stage process of progressive improvement in the level of detail and accuracy. Inevitably the process will include a number of iterations between the different stages and can be tailored to fit in with Gateway reviews such as the OGC Gateway™ Process (www.ogc.gov.uk). Box 2 illustrates an indicative process that can be used to develop such solutions.

Achieving value from contract strategies – early supply chain involvement is important

Table 2 summarises different contract strategies and situations where they are most likely to be helpful.

Traditional contracting versus a partnering approach

Traditional contractual arrangements for the supply of constructed assets involve a contract established between a client and the main contractor. There are many different forms of contract, but they all set out what is to be provided and the fee to be paid. In the UK, most construction professionals who design carry professional indemnity insurance; however, contractors do not normally carry this insurance. It follows that contractors are only insured if they follow to the letter the instructions of the design team. All these instructions are contained in a document called the design specification, supported by design drawings.

Box 2 Stages of the WLV process

1. Inception

A strategic business case can be made to justify the project and approximate budget prices (based on coarse whole life cost analysis) are identified. The following issues are considered:

- Identify value drivers/criteria – stakeholders and their requirements.
- Define the need in terms of functional requirements.
- Consider constraints.
- Develop value measures for an appraisal of options.
- Carry out a high level assessment of risks and opportunities.
- Consider alternatives, eg between leasing, buying or building or repairing/adapting existing asset.
- Define outcome requirements eg rate of return on capital investment, time to achieve occupation, environmental impact permissible.
- Confirm funding source and identify possible procurement routes.

2. Feasibility appraisals

Alternative options are developed and analysed to determine their feasibility and ensure that they align with the outcome requirements. These will form part of an outline business case, which can then be developed into a more fully defined business case. Relevant issues are:

- Develop alternative options.
- Define analysis criteria (eg design life, type of use).
- Evaluate risks and identify means to manage the risks.
- Allow for the effects of Optimism Bias (OB), eg underestimation of costs and over estimation of benefits.
- Develop and use decision support tools, eg WLC/LCA – bespoke or off-the shelf.
- Compare and analyse the options, and select the preferred option.
- Confirm the selected procurement method.
- Identify the best value solution.

3. Plan and design

The detailed design is developed and will involve input from designers, planners, architects working closely with supply chain, contractors, suppliers and manufacturers:

- Refine the whole-life cost plans.
- Replace original assumptions with better assessments eg quantities, price and predicted; performance of alternative components, materials and services.
- Carry out detailed assessments of risks and uncertainties.
- Develop whole life plans supported by durability information and planned maintenance activity profiles.
- Progressively replace historic cost estimates with predicted costs.
- Consider health and safety management aspects, eg the Construction Design Management Regulations.
- Refine timing of costs and benefits.
- Select optimum option by measuring against outcome specification.

4. Construct and handover

- Develop detailed work plans and timetable for completion.
- Work with stakeholders as appropriate, eg inform them about any possible disruptions.
- Audit to confirm compliance with specification.
- Handover including documentation on operation in use, including Construction Design Management (CDM) requirements for health and safety in use.

5. Operate and maintain

- Operate and maintain to specified standards.
- Monitor in-service performance (identify deviations from plans in respect of durability and timing of maintenance activities).
- Develop feedback loops to enable future designs to be modified.

6. Decommission and renewal

- Consider way forward – demolish and reconstruct or upgrade.
- Dispose for re-use by a new owner.
- Consider income and/or costs.
- Consider environmental impacts, disposal, or re-use of materials.

Table 2 Common contract strategies in procurement

Contract strategy	Situation where it is most helpful
Limited competition with individual price negotiation	Requirement for low value, low volume goods and services from a known good supplier in a niche market. Tender costs are disproportionate to the value of the contract and competition is unlikely to reveal other potential suppliers
Competitive tender (depending on value for public sector these must be advertised in the *Official Journal of the European Communities*)	Requirement for medium to high volume with more than one potential supplier available
Framework agreements (like call-off contracts, but there are a number of suppliers to choose from, identified through competition)	Requirement for several specific goods or services within a broad category, over a specified period, but exact requirement cannot be predicted
Lump sum contracts	Usually used for purchasing professional services where the exact quantity requirement is unknown
PFI/PPPs	Where cost and quality of delivery of services can be improved by use of private sector management and business skills incentivised by having private finance at risk and by transferring risks to those best able to manage them

Source: Adapted from *Getting Value for Money from Procurement: How auditors can help*. (NAO/OGC, 2001)

The traditional contracting approach specifically excludes the contractor/builder from taking any active part in construction design. Any mistakes made by the designers due to lack of experience, or lack of knowledge, inevitably hamper the work of the contractors.

Partnering initiatives concentrate on the need for, and the mechanisms for, developing better client/supplier relationships in order to deliver better business value through a project's life.

Advice on how to develop a partnering approach to contracting in construction is available from authorities including the OGC website (www.ogc.gov.uk). The Audit Commission has produced a bulletin *Developing productive partnerships* (Audit Commission website www.district-audit.gov.uk) including a list of references covering just about everything you need to know about partnerships and partnership working.

Major clients drive early supply chain involvement

The Defence Estates piloted Building Down Barriers (BDB) as a systematic and managed approach to procurement and maintenance of buildings based on a pre-assembled supply chain, involving major contractors early in the project development process. This ensures a single point of responsibility from a prime contractor to deliver the outputs using integrated teams of client, contractor, specialist contractors and suppliers. The lessons learned were published by CIRIA in the *Handbook of supply chain management* (Holti *et al*., 2000).

The Ministry of Defence is one of the largest landowners in the UK and currently spends over £1bn per annum on its estate, which comprises some 240 000 hectares with over 4000 sites. Sites can be broadly described as 'built' (barracks, naval bases, depots, aircraft hangars, etc.) or 'rural'. The Defence Estates is an executive agency of the Ministry of Defence that ensures that the estate is managed and developed in a sustainable manner, in line with acknowledged best practice and government policy.

In the private sector, clients such as the British Airports Authority (BAA) have developed their own project process (British Airports Authority, 1996). (see *Design Chains – a handbook for integrated collaborative design* [Austin *et al*, 2001] funded by Amec, the DTI and the Engineering and Physical Sciences Research Council [EPSRC]).

section 7
Techniques and methods

summary
Users will benefit from decision-making tools that provide a clear understanding of all the factors to carry out comprehensive relative assessments of WLV

The key techniques that are integral to WLV evaluations of building and infrastructure projects include:

- WLC and LCA: WLC deals primarily (but not exclusively) with financial costs, whereas LCA deals primarily (but not exclusively) with environmental impacts. Individually, WLC and LCA techniques cannot comprehensively cover all financial, environmental, and social costs and benefits associated with achieving the best WLV, and assessing alternative options.
- MCA: MCA is used in conjunction with both WLC and LCA to evaluate alternative options based on criteria developed with stakeholders.
- Group decision-making processes: These processes include VM and RM processes to engage stakeholder participation for achieving WLV.

Quantitative assessment is often easier to undertake than qualitative assessment but ultimately the most successful decision support tools may have to assess both quantitative and qualitative factors.

Whole Life Costing

BS ISO 15686-1 defines WLC as an:

"economic assessment considering all agreed projected significant and relevant cost flows over a period of analysis expressed in monetary value. The projected costs are those needed to achieve defined levels of performance, including reliability, safety and availability".

It is a process that aims to look at every cost incurred in respect of a facility or product from inception to disposal, ie the total costs associated with the procurement, use during service-life, and disposal at the end of life.

The objective of WLC is to make investment decisions with a full understanding of the cost consequences of different initial options. The technique is not new and has been applied to some extent in different sectors recently, as illustrated in the case studies (Section 10). Over the last few years the term WLC has come to mean different things to different people. These can be separated into 'simple' and 'complex' uses and meanings.

- Simple uses of WLC limit the word 'costing' to financial impacts.
- Complex uses extend 'costing' to include some external effects, eg impacts on the environment, users and society in general, by representing these in financial terms.

WLC can be used at any stage in the life of an asset (new build or maintenance and refurbishment) and also by all organisations involved in the stages of the supply chain. At each stage activities required are identified, together with their probable costing and timing. The cash flows resulting from the analysis are then brought back to a common basis of measurement, normally by using discounted cash flow techniques. This entails using a discount rate to reflect the present value of the whole cash flow analysis.

Net Present Value (NPV), a measure often used in WLC, is the sum of the discounted benefits of an option less the sum of the discounted costs. It represents therefore a single figure, which takes account of all relevant future incomes and expenditures for that option over the period of analysis. The discounted cash flow analysis gives a basis for comparison between the WLC of each option.

Normally this calculation will be undertaken on either a spreadsheet program (such as MS Excel) or on a specialist WLC program. Specialist programs will normally integrate functions such as sensitivity analysis of the inputs and/or default or benchmark data inputs. An example of an option comparison using WLC is shown in Table 3 and Figure 7.

As projects progress, initial assumptions will be substituted by real projections based on the project design, project programme and specification, together with analysis of probable deterioration over time and associated requirements to undertake maintenance activities. The technical analysis for these inputs will normally be made by professionals within the project team, or by specialists in whole life performance.

Typical breakdown of the costs involved in the three main stages of WLC are shown in Table 4.

WLC is not an exact science and the technique is therefore often used in conjunction with other processes such as RM, VM and Value Engineering (VE). For example, for most facilities, it is difficult to identify 'end of life' so

Table 3 WLC option comparison

Option 1	Year 1	Year 2	Year 3	Year 4	Year 5
Decorations total (£)	10 000	10 000	10 000	10 000	10 000
Fabric total (£)	8 769	8 769	8 769	8 769	30 000
Services installations total (£)	200	200	200	200	200
External work total (£)	1 500	1 500	3 500	3 500	3 500
Subtotals before discount (£)	***20 469***	***20 469***	***22 469***	***22 469***	***43 700***
Discounting rate of 3.5 % per annum	1	0.966184	0.933511	0.901943	0.871442
NPV (£)	20 469	19 777	20 975	20 266	38 082
Cumulative NPV (£)	**20 469**	**40 246**	**61 221**	**81 487**	**119 569**
Option 2	**Year 1**	**Year 2**	**Year 3**	**Year 4**	**Year 5**
Decorations total (£)	10 000	10 000	0	30 000	20 000
Fabric total (£)	0	0	18 000	18 000	28 000
Services installations total (£)	500	500	500	500	1 500
External work total (£)	1 500	1 500	1 500	1 500	5 000
Subtotals before discount (£)	***12 000***	***12 000***	***20 000***	***50 000***	***54 500***
Discounting rate of 3.5 % per annum	1	0.966184	0.933511	0.901943	0.871442
NPV (£)	12 000	11 594	18 670	45 097	47 494
Cumulative NPV (£)	**12 000**	**23 594**	**42 264**	**87 362**	**134 855**

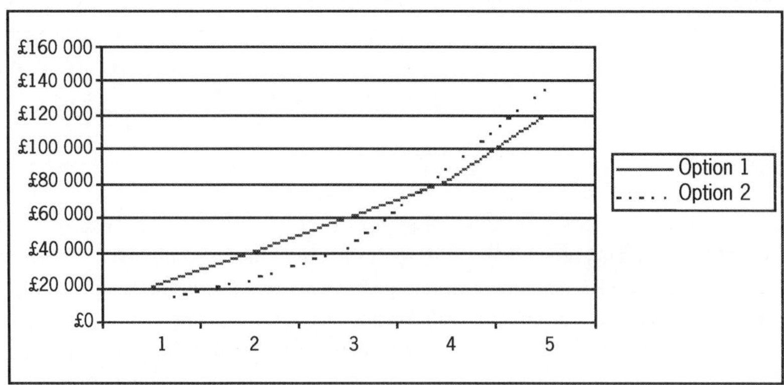

Figure 7 A WLC option comparison

that it is not just a question of identifying what the asset will cost over a period of time but also identifying the appropriate time horizon for the analysis. Evaluations are often based on a fixed 'life' period that takes account of major cost consequences of the initial choice or contractual responsibilities for funding maintenance, eg 40 years for road pavements or 25 years for typical PFI projects.

The discount rate used in the analysis has a major effect on the NPV of the asset. High discount rates favour assets with a high expenditure towards the end of the life of the asset while low discount rates favour assets with high initial costs and low in-service costs. Most often, and particularly for public sector procurement, clients may not have the option of setting the discount rate. For example, for transport infrastructure procurement, HM Treasury fixes the discount rate at 3.5% over the first 30 years. In the private sector however, the discount rate is base on the financial markets and the expected rate of return. The choice of discount rate also influences the selection of a realistic analysis period, with longer evaluation

Table 4 WLC breakdown

Acquisition	Use and maintenance	Disposal
Costs ● Acquisition by construction ● Site costs eg purchase, clearance and groundwork ● Design ● Preliminaries ● Construction commissioning and/or fitting out ● Fees ● In-house administration	***Costs*** ● Operational costs ● Routine maintenance, eg cleaning, energy, utilities, facilities management including landscape maintenance ● Major maintenance, commissioning and/or fitting out, eg repairs, churn costs	***Costs*** ● Cost of disposal eg demolition, sale ● Site clean-up
Acquisition by purchase/leasing ● Purchase price ● Cost of purchase/adaptation ● Fees ● In-house administration	***Income*** ● Income generated through ownership of asset eg rents from surplus space less loss of income during refurbishment or during failure of facilities	***Income*** ● Sale or interest in asset ● Sale of materials for reuse/recycle

Source: *Whole Life Costing – A Client's Guide* (CCF, 1999)

periods being supported by low discount rates.

To enable a fair comparison of alternative products or options with different patterns of spend, the residual value of the asset at the end of the analysis period is another key element of the process.

Typical decisions informed by WLC analysis include:

- Choices between alternative components, all of which have acceptable performance (component level WLC analysis).
- Choices between alternative designs for the whole, or part, of a constructed asset (assembly or whole asset level WLC analysis).
- Evaluation of different investment scenarios, eg to adapt and redevelop an existing facility, or to provide a totally new facility.
- Estimation of future costs for budgets or the evaluation of the acceptability of an investment on the basis of cost of ownership.
- Comparison and/or benchmarking analysis of previous investment decisions, which may be at the level of individual cost headings, eg energy costs, cleaning costs, or at a strategic level, eg open plan versus cellular office accommodation.

Life Cycle Assessments

ISO 14040 defines LCA as:

> "a systematic set of procedures for compiling and examining the inputs and outputs of materials and energy and the associated environmental impacts directly attributable to the functioning of a product or service system throughout its life cycle".

LCA studies consist of four interrelated stages as shown in Figure 8, and guidance on how to conduct each stage is given in the ISO 14000 series of international standards.

A 'full' LCA describes all the interactions between the construction system under study and the environment over the whole life-cycle of a product or service. However, in practice this is often either not possible or is impractical because of the lack of data. As a consequence, LCA studies require careful attention to setting boundaries between the construction system under study and the environment, so that the information generated is fit for purpose. This includes setting boundaries, for example:

- Geographical boundaries. Establishing if the study is for a whole country or one building.
- Temporal boundaries. By setting the time period over which the environmental impacts will be assessed.

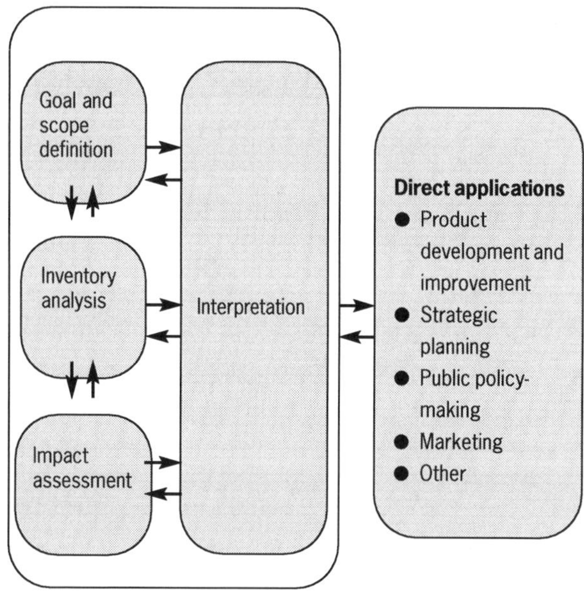

Figure 8 Stages in conducting LCA studies (adapted from BS EN ISO 14040)

Example of LCA assessment of bridge maintenance strategies

Who?

Surrey County Council (SCC)

Why?

SCC needed to develop a system to evaluate the environmental performance of selected bridge management strategies as a decision support tool. This would enable the SCC to meet its obligations under Agenda 21 (best value and continuous improvement in reducing environmental impacts).

How?

A four stage assessment process was adopted:

1. Structural classification (beam, arch or cable).
2. Codify service life management options (deterioration problems and faults, and applicable maintenance, eg preventative, remedial, strengthening, structural alteration, accident repair).
3. Develop idealised life-cycles of bridge management practice (by case study investigation and questionnaires).
4. LCA of each type of bridge, ie masonry/brick arch type, reinforced concrete types and metal types, as illustrated in Figure 9.

The next phase was to use these idealised analyses to inform actual decision-making on individual bridges.

- Process boundaries. Deciding to include or exclude processes like maintenance from the assessment.

Strategic LCA approaches

Strategic LCA approaches can be used to provide high-level advice directly to government or to strategic planning departments of private organisations. Strictly speaking, such methods are not 'true' LCAs within the meaning of ISO 14040. However, they do provide general information about material and energy flows in the economy, usually based on national statistics, eg environmental accounts information from the Office of National Statistics in the UK.

Conventional LCA approaches

Conventional LCA approaches are those that fall within the practices described in ISO 14040. They are used for evaluating construction products and services at national and international levels. At the time of writing, they are also contributing to the rapidly evolving field of Environmental Product Declarations (EPDs) for construction products, and it is expected that they will be incorporated, in some form, into EU Construction Product Declarations (CPDs) in the near future. This is perhaps the single largest driver for the rapidly growing interest in applying LCA to evaluate construction products. These approaches can generally be regarded as streamlined approaches to LCA, in that:

- The number of environmental parameters in the analysis may be limited to those of greatest interest to decision makers, which reduces the data required for the analysis.

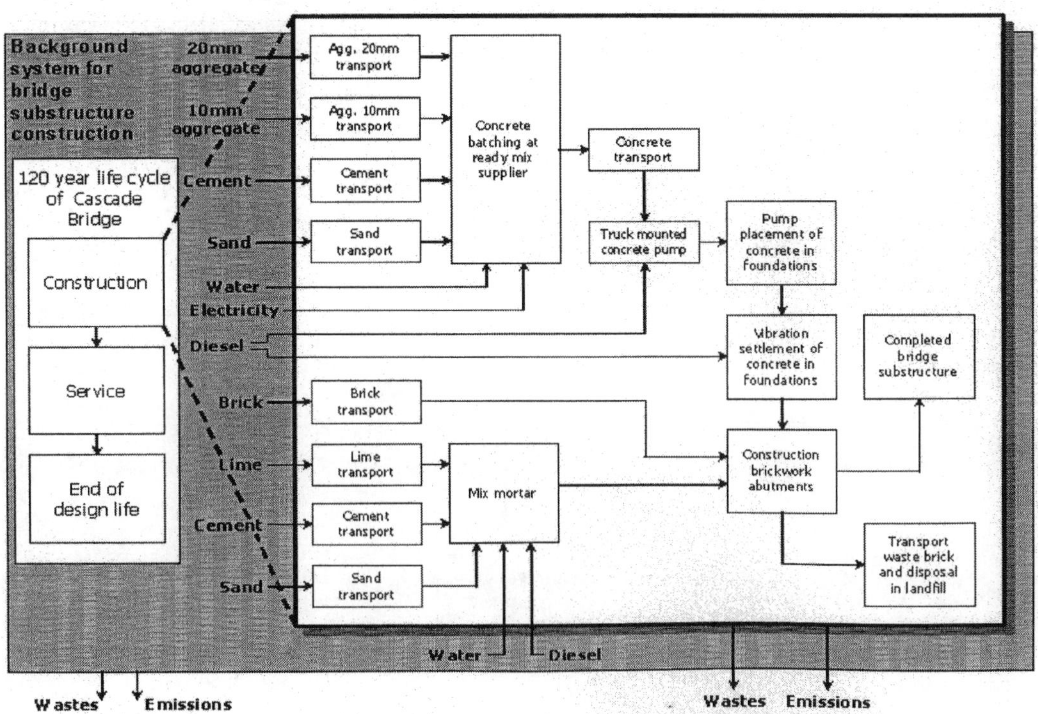

Figure 9 LCA example for a bridge

- They can reduce the number of life-cycle stages covered by the analysis.
- They need not necessarily include a Life Cycle Impact Assessment (LCIA) phase, ie express results as emissions not impacts on the environment, eg generating results as tonnes of carbon dioxide emissions rather than describing global warming potential.
- They are simple enough to allow laypersons to interpret the results after minimal training.

However, whenever LCA approaches are employed, the context in which the results will be used should be evaluated carefully. Considerations include (The Society of Environmental Toxicology and Chemistry [SETAC], 1999):

- How will the results be used?
- Is there a dominant life-cycle stage?
- What is the threshold of uncertainty?
- How narrowly is the product defined (generic or specific)?
- How much is already known about the life-cycle stages of the product?

If any of the following points apply to the construction product/services, a full LCA should be considered:

- The results are to be used for public policy applications or other uses in the public sphere (eg marketing).
- There is no dominant life-cycle stage.
- The data are uncertain.
- The product is specific.
- Little is known about the life-cycle stages of the product.

This information can be used to provide simpler tools to the industry, concentrating on those life-cycle steps and resource inputs or emissions that cause the greatest environmental problems. Conversely, if only a generic assessment of potential products to use in a project is required and the life-cycle characteristics are well known, it is legitimate to streamline assessments where the results are not disclosed publicly.

Integrating Life Cycle Assessment with Whole Life Costing

The integration of WLC and LCA presents a powerful route to obtain 'best value' solutions in both financial

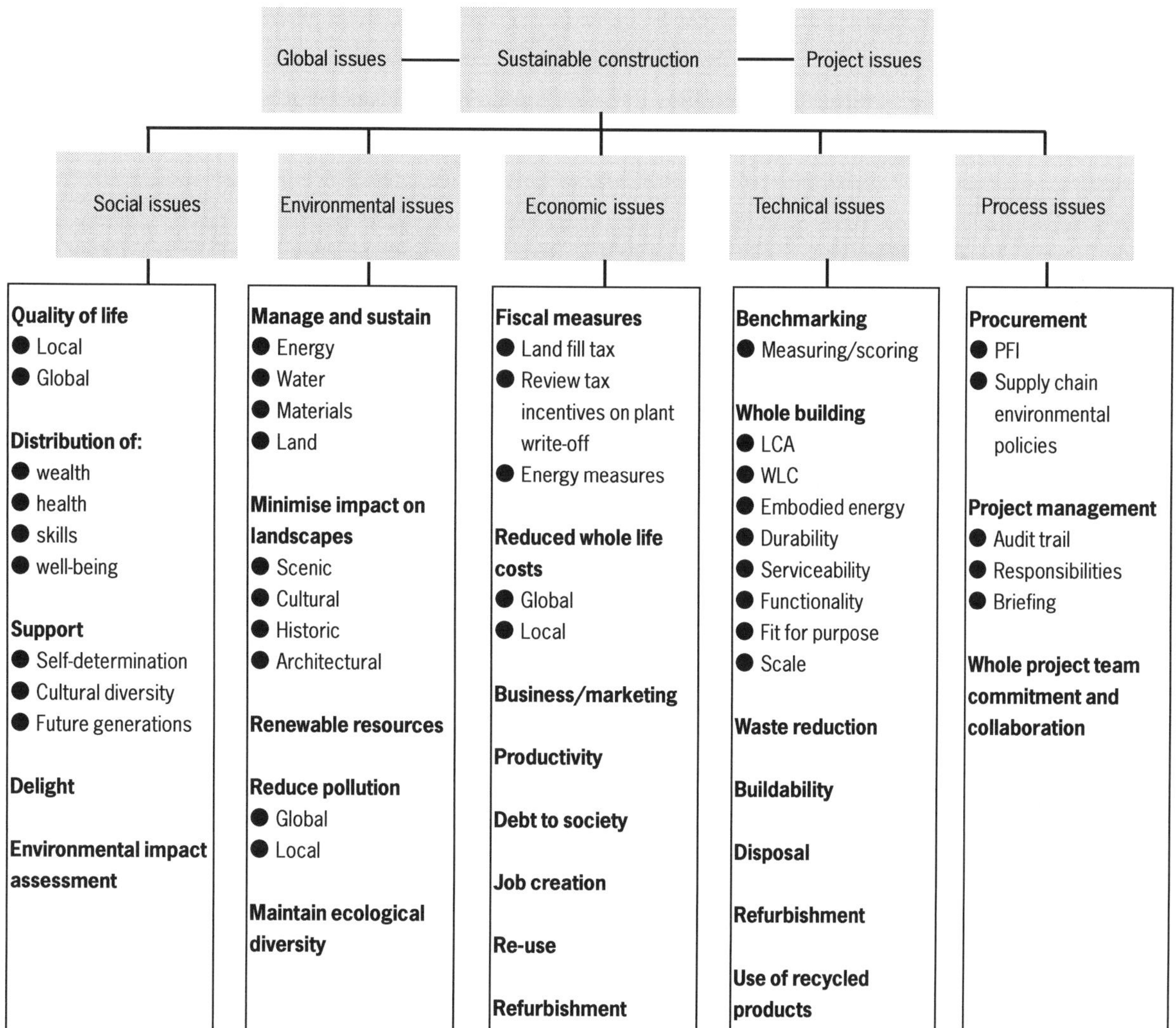

Figure 10 Issues considered by WLC and LCA

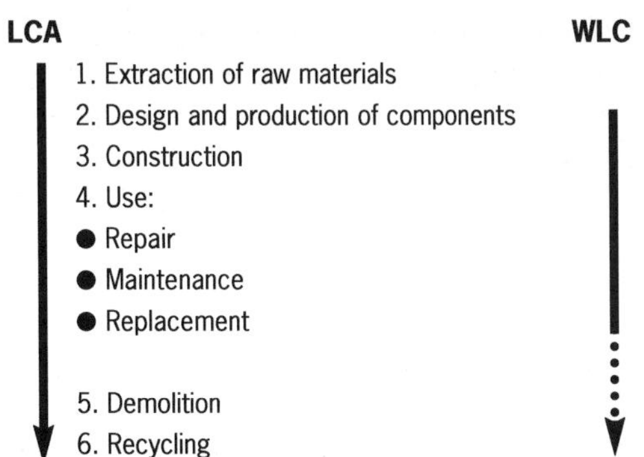

Figure 11 Commonality within the life-cycle of issues relating to WLC and LCA

and environmental terms and has the potential to make a significant contribution to achieving sustainable development.

There is commonality in some of the issues considered in both techniques (Figure 10), in addition to both techniques addressing the life-cycle (see Figure 11). This can lead to synergies in the analysis, but also potentially to conflicts between the conclusions resulting from use of each technique. Generally these will be minimised when the techniques are used jointly from the outset on a project.

Multi-Criteria Analysis

MCA assists in prioritising options from the most preferred to the least preferred option. It has much potential for construction as it provides a clear audit trail to decisions and procurement processes by using criteria scoring and weighting, eg to appraise tender bids and vendors.

MCA is based on simple principles of preference: if A is preferred to B, and B to C, then A should be preferred to C. MCA helps unlock WLV by making decisions that are clear, unambiguous and justifiable – the crucial factors in public evaluation, accountability and decision-making. It enables participatory decision-making by allowing stakeholders to take part in the decision-making process by either shaping, and/or transforming their preferences, or making decisions in conformity with identified goals set.

MCA is conducted in workshops involving all interest groups and key players with an impartial facilitator guiding the group to help to:

- prioritise options.
- identify areas of opportunity.
- clarify the differences between options.
- help key players to understand the situation better.
- indicate the best allocation of resources to achieve the goals.
- facilitate the generation of new and better options.
- improve communication between stakeholders.
- decide the best way forward.

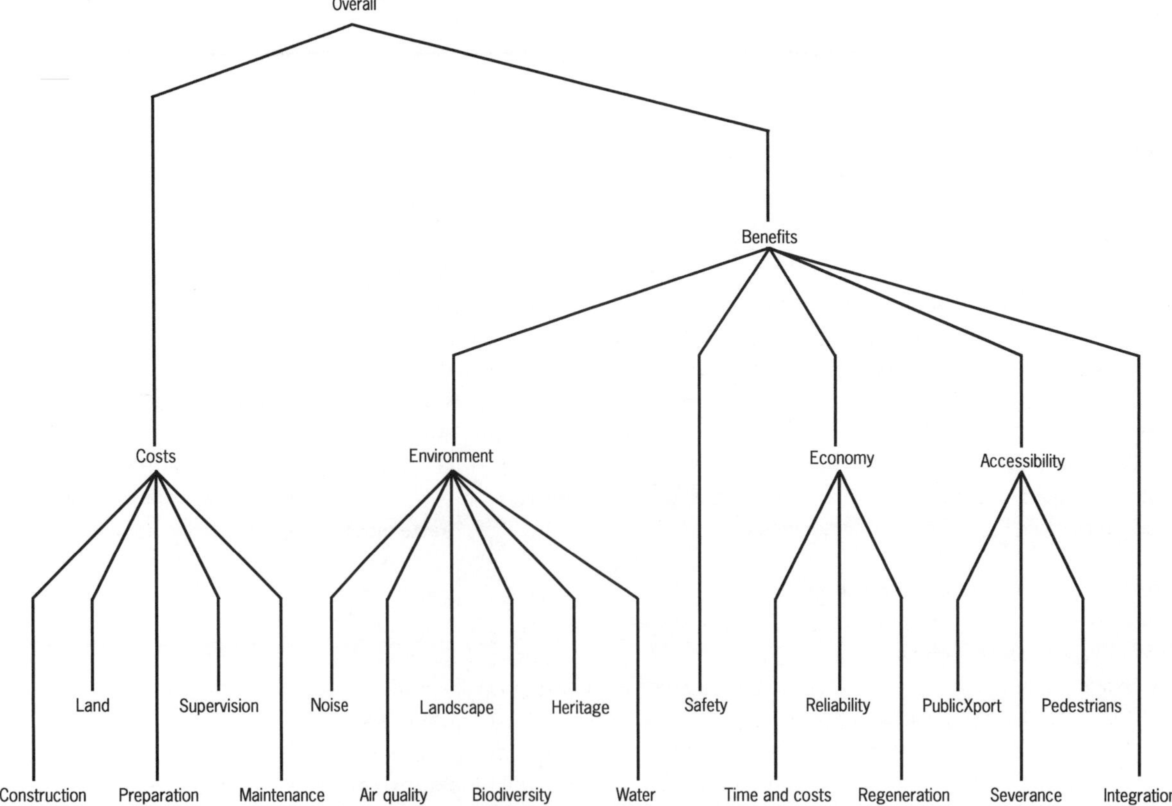

Figure 12 A value tree of objectives (Source: DETR MCA Manual)

Figure 12 shows how objectives and criteria for the DETR's approach to appraisal of transport investments might be represented. Organising the objectives and criteria in a value tree (Figure 12) often highlights the conflict amongst the objectives, and this can lead to refining their definitions as well as stimulate thinking about new options that could reduce the apparent conflicts between the objectives.

MCA methods differ from LCA and WLC mainly in the breadth of different decision criteria considered in the methodology and the transparency of its output, which makes use of qualitative and quantitative data inputs. For both reasons, it has proved useful in policy decisions where there are many options to be appraised using a variety of criteria and provides a transparent framework for exploring where trade-offs between objectives occur in the decision-making process.

The results for key criteria used in LCA and WLC assessments for a set of decision options can be incorporated into an MCA framework as data input to score the options. For example, the global warming potential over the life-cycle of a product calculated in an LCA study could be used as data input to an MCA to decide which of a number of products to buy; its cost could also be used as a separate criterion. This would provide a robust valuation of the performance of the options against a number of criteria, rather than focussing on environmental or cost criteria exclusively.

The use of an MCA framework would allow the inclusion of social decision criteria in the assessment, which is central to the increasing emphasis on addressing sustainable development objectives in decision-making activities. This requires an appraisal against social, environmental and economic objectives, which is not possible using either LCA or WLC alone.

MCA also provides an audit trail for the decision-making process, so that if the outcome is questioned at a later date, there is evidence available to the auditor to demonstrate how a decision was arrived at. This is a useful feature for decisions that could have significant environmental effects associated with them, eg siting new quarries, given that the increasing public interest implies that the process will be subject to increasing scrutiny in future. For example, the increasing number of public enquiries and judicial reviews arising from such decisions, which require better documentation describing the deliberation process and its outcome. This is addressed in a more comprehensive and transparent manner by MCA methods than LCA or WLC methods.

MCA is therefore useful in such policy decision contexts, especially where monetary valuations are not generally regarded as very robust, eg for describing the consequences of environmental impacts, or are difficult to use because there are no agreed methods of valuation for key decision criteria. In these cases, decision makers might consider either the use of sensitivity assessment to see how much results depend on the particular values used, or ignore monetary values and rely on subjective scoring and weighting systems to reflect decision makers' or interest groups' preferences.

MCA: an example from transport infrastructure

Cost benefit analysis has been used to appraise road investment in the UK from the 1960s onwards. However, with increasing awareness of the environmental consequences of major transport investment schemes, the Department for Transport has used MCA, defined as:

> "a process that enables the identification of preferences between options by reference to an explicit set of objectives that the decision making body has identified, and for which it has established measurable criteria to assess the extent to which the objectives have been achieved".

This allowed the Department of Transport to develop a framework for considering monetary and non-monetary information for the appraisal for road and other transport infrastructure projects. An approach to incorporate non-monetary elements into the decision framework by measuring them on numerical scales or by including qualitative descriptions of the effects, in an Environmental Impact Statement was formalised in 1998.

Guidance on the Methodology for Multi-Modal Studies published in 2000 by the Department sets out how an Appraisal Summary Table (AST) may be used to summarise the main economic, environmental and social impacts of each transport option considered in a decision-making process as a single table. The AST contains the scores for the option against the five policy objectives requiring evaluation in the decision-making process, and their sub-objectives, as listed below:

Environment objective

- Protect the built and natural environment
- Reduce noise
- Improve local air quality
- Reduce greenhouse gases
- Protect and enhance the landscape
- Protect and enhance the townscape
- Protect the heritage of historic resources
- Support biodiversity
- Protect the water environment
- Encourage physical fitness
- Improve journey ambience.

Safety objective

- Improve safety
- Reduce accidents
- Improve security.

Economy objective

- Support sustainable economic activity and get good value for money
- Improve transport economic efficiency

Table 5 AST 1 – Summary impacts of baseline option (2.0 and 2.1 km cut and cover scheme) compared to 'Do Minimum' option

Objectives	Permanent impacts in operation
Environmental	
Noise	Large beneficial
Local air quality	Neutral
Construction dust	N/A
Landscape	Moderate beneficial
Cultural heritage	Large beneficial
Biodiversity	Large beneficial
Water	Slight adverse
Agriculture	Moderate beneficial
Journey ambience	Moderate adverse
Visitor amenity	Moderate beneficial
Non-motorised users	Moderate beneficial
Environmental resource	
Use of non-renewable resources	N/A
Energy usage	Neutral
Other waste	N/A
Decommissioning	Moderate adverse
Transport of goods and services (site)	N/A
Longevity	No suitable baseline comparison

- Improve reliability
- Provide beneficial wider economic impacts.

Accessibility objective

- Improve access to facilities for those without a car and to reduce severance
- Improve access to the transport system
- Increase option values
- Reduce severance.

Integration objective

- Ensure that all decisions are taken in the context of the Government's integrated transport policy
- Improve transport interchange
- Integrate transport policy with land-use policy
- Integrate transport policy with other government policies.

ASTs for A303 Stonehenge Improvement project

The ASTs for the appraisal of the environmental objectives of alternative options for the A303 Stonehenge Improvement project are shown in Tables 5 and 6 to illustrate the technique. This is an example of an application of an MCA framework for decision support that does not use any form of weighting. The ASTs compare different options for improving a road or building a tunnel or 'Do Minimum'.

Table 6 AST 2 – Impacts of alternative bored tunnel options relative to 2.1 km 'cut and cover baseline' option

Objectives	Permanent impacts in operation			Construction impacts		
	2.1 km	2.67 km	4.53 km	2.1 km	2.67 km	4.53 km
Environmental						
Noise	SB	MB		MB	MB	
Local air quality	N	N		N/A	N/A	
Construction dust	N/A	N/A		SA	SA	
Landscape	MB	MB		MB	SB	
Cultural heritage	SB	MB	LB	SB	MB	
Biodiversity	SB	SB		SB	SB	SA
Water	S/MA	S/MA	MA	S/MA	S/MA	MA
Agriculture	N	SB	LB	MB	SB	MB
Journey ambience	N	SA	LA	MB	SA	MA
Visitor amenity	SB	MB	SB	MB	SB	MB
Non-motorised users	N	N		SB	SB	
Environmental resource						
Use of non-renewable resources	N/A	N/A		MB	SB	MA
Energy usage (construction)	MA	MA	LA	MA	MA	LA
Other waste	MA	MA		MA	MA	
Decommissioning	MB	N	SA	N/A	N/A	
Transport of goods and services (site)	N/A	N/A		SB	SB	
Longevity	SB	SB		N/A	N/A	

Key: SB = slight beneficial; MB = moderate beneficial; LB = large beneficial; MA = moderate adverse; SA = slight adverse; MA = moderate adverse; LA = large adverse; S/MA = slight/moderate adverse; N = neutral

Source: Stonehenge WHS Management Plan, Highways Agency. Available from: http://www.highways.gov.uk/roads/projects/a_roads/a303/stonehenge/index.htm

Risk Management and Value Management processes

Risk assessment and RM is the systematic process of identifying and managing risks and opportunities for a project or business. It includes the following stages:

- Identify risks and opportunities by brainstorming or other creative techniques.
- Assess their probabilities of occurrence and consequences through:
  – qualitative approach involving simple scales for prioritising risks, and/or
  – quantitative approach involving modelling and simulation, taking account of Optimism Bias (OB).
- Develop a risk mitigation and contingency plan.
- Allocate risk ownership to the party which is best able to bear the risk.
- Participation of stakeholders involved.

The Green Book (HM Treasury, 2003) requires that OB be taken into account when assessing the costs of risks at the early stages of project development. OB takes account of a systematic tendency by project appraisers to be over optimistic when estimating benefits and tend to understate timings and costs, both capital and operational. To redress this tendency, appraisers should make adjustments by increasing the cost estimates and decreasing, and delaying the receipt of, estimated benefits. *The Green Book* also recommends that sensitivity analysis be used to test assumptions about operating costs and expected benefits.

The likely deliverables of such a RM process include, depending on the depth of the RM study:

- A detailed risk register, including all risks, probabilities, consequences, cost impacts, mitigation actions etc. Most important risks are known.
- Mitigation actions are developed for important risks.
- Risk contingency estimates are determined with reasonable certainty.
- Risk allowance for a project's costs.
- A decision aid for option selection:
  – Decisions made are based on increased certainty over future events.
  – Contingency costs and measures will have been assessed and justified by project stakeholders.
  – More certainty for implementing the project.

Many techniques are available for identifying, assessing and managing risks throughout a project's life. Table 7 provides an overview of these techniques and their relative suitability throughout a project's life-cycle.

VM is a structured, systematic and participatory decision-making process to develop best-value optimum solutions, to ensure client's objectives are achieved with stakeholders' buy-in. VM evolved from Value Engineering (VE), which originated during the Second World War at General Electric and is nowadays a proven management process used internationally in many industries such as construction, services, government, and manufacturing.

BS EN 12973: 2000 defines VM as:

"a style of management, particularly dedicated to motivating people, developing skills and promoting synergies and innovation, with the aim of maximising the overall performance of an organisation. Applied at the corporate level, value management relies on a value-based organisational culture taking into account value for both stakeholders and customers. At the operational level (project oriented activities) it implies, in addition, the use of appropriate methods and tools".

Table 7 Risk management techniques – when to use them in the project life-cycle

	Project life-cycle phase					
Risk techniques	**Inception**	**Feasibility appraisals**	**Plan and design**	**Construct and handover**	**Operate and maintain**	**Decommission and renewal**
Brainstorming	G	G	F	F	P	F
Checklists	P	P	G	G	G	G
Prompt lists	G	G	N	N	N	G
Assumptions analysis	G	G	F	N	N	G
Delphi technique	G	P	N	N	N	G
Interviews	P	G	G	G	P	F
Risk register	P	G	G	G	P	F
Probability-impact tables	G	G	G	G	P	F
Decision trees	G	F	P	P	N	F
Influence diagrams	G	F	F	F	N	F
Monte Carlo simulation	N	G	G	G	N	G
Sensitivity analysis	F	G	P	P	N	P

Key: G = Good (relatively strong application; F = Fair (average application; P = Poor (relatively weak application; N = None (not applicable))
Source: Adapted from *Project Risk Analysis Guide*, Association for Project management, 1997.

In the construction industry context, VM can be defined as:

"a structured approach to defining what value means to a client in meeting a perceived need by establishing a clear consensus about the project objectives and how they can be achieved". (CIRIA SP129, 1996)

VE is incorporated into VM as a systematic approach to delivering the required functions at lowest cost without detriment to quality, performance and reliability (CIRIA SP129, 1996). Function analysis is one of the fundamental techniques involved in a VE study. Its purpose is to develop a systematic breakdown of functional requirements, concentrating on the actual needs, aspirations and wants of the client and project stakeholders.

A VM study can be applied at varying stages of a project but for best results, a study should be done early in the planning/concept stage to ensure that the best project strategy is adopted, to reduce uncertainty and identify future opportunities. The process should involve all the relevant stakeholders sitting in the same room and making informed decisions in a rational way and achieve consensus under the guidance of an expert facilitator through successive collaborative workshops. An example of a VM process is illustrated in Section 4.

Integrated Risk Management and Value Management

In recent years, methodologies for the integration of value and RM have been developed for understanding, optimising, communicating and tracking the different perspectives of value and risk throughout all project development phases. Figure 13 shows the application of integrated risk and VM in PPPs (CIRIA C617, 2003).

Figure 13 illustrates the development of service values, their judged importance, the risks associated with achieving each of the values and how these risks are captured in risk registers. This knowledge is further refined through successive iterations and communicated to the PPP project stakeholders throughout the project life-cycle.

Integration of value assessments with outputs of tools

Approaches to delivering WLV need to help each stakeholder make structured assessments of the value (to them) of the built product and the services it supports. These assessments can be shared by the project team during design and construction to help guide their decision-making process. Value can be usefully expressed as the trade off between the sacrifices a person or organisation is willing to make in order to receive benefits from using or owning a product or from being involved in its production. In simple terms, value is about what you get for what you give up. This deliberately broad definition of value avoids focusing on cost alone. It allows WLV to be considered in a way that helps stakeholders become confident that they are 'doing the right thing.' This, in turn, helps them make key design decisions that have a knock-on effect on through-life performance. The benefits and sacrifices view of WLV also allows performance targets to be set. These targets describe the extent of the sacrifices stakeholders are willing to make and the extent of the benefits they seek to experience throughout the product life in return.

The VALiD Framework (see www.valueindesign.com) helps project stakeholders discuss value by helping them understand the influence of their values over their judgements of WLV. In this framework, value is represented by the stakeholders' collective judgements of the extent to which the entire project's benefits and sacrifices are represented, compared with the targets set.

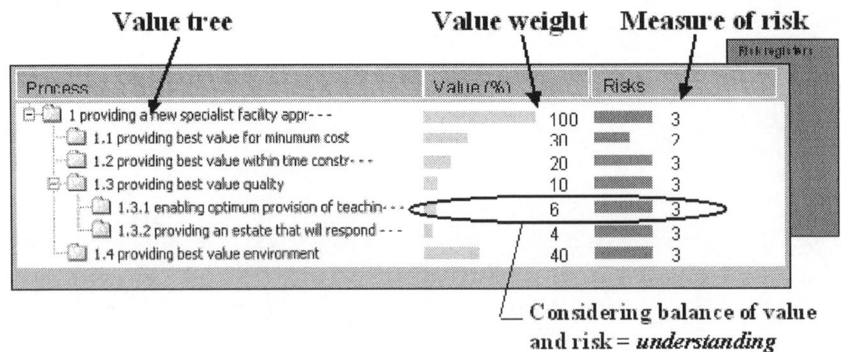

Figure 13 Value and risk management in PPP. Source: PPPCOM *Toolkit for managing risk and Value in PPP* (CIRIA C617, 2003)

section 8
Tools and resources

summary

Many tools are available in the public domain which provide practical advice and guidance on all aspects of WLV

There are numerous tools in the public domain that may help you evaluate WLV for particular projects, depending on the type of project, who is undertaking the evaluation, the stage of the project and the focus of the evaluation. We have selected a short list of the most relevant tools with brief descriptions and links which is attached in Annex 2. The case studies in section 10 also illustrates the use of several of these tools in more detail and include reference points for further details.

You may also find that the descriptions and tools provide key words and relevant websites for searching the Internet. The majority of the tools and resources summarised in Annex 2 (and other guidance) are included on the WLV website (www.wlv.org.uk) and are available to use free of charge. The free search facility on this website (which originated from a DTI sponsored project in 2001 – 2003 on Designing for Whole Life Value) allows you to identify tools with a social, economic or environmental angle on WLV. It also includes links to access relevant websites for each tool.

section 9
Future challenges

In the course of writing this guide, we have identified various barriers and challenges implicit in moving to WLV as the basis of an appraisal of options. Efficient and effective application of WLV principles will need to take account of all these issues in the future. The WLV approach covered in this guide will provide you with reasonable confidence to consider these issues in achieving WLV in infrastructure and buildings projects.

This guide is a step on the way, but there are still some challenges to be addressed. The following is a summary of some of these challenges.

Consistency of application

How do you make sure that WLV principles are applied consistently in the development and management of assets throughout? Do we need more incentive and policy initiatives to encourage their applications? For example an appraisal of options, based on a model integrating WLV and sustainability, would need more data and take longer than is currently the case. How motivated and committed are clients to carry such comprehensive assessments?

Reliability

How do you ensure that data used at the various stages are available and reliable? WLV data management systems or asset information systems need to be strengthened to ensure that reliable data, eg assets' service lives, on maintenance and renewal regimes etc, is collected on infrastructure and building assets. These systems must be able to share relevant data for WLC and LCA between relevant stakeholders so that informed decisions can be made and decisions are based on a WLV approach.

Stakeholder involvement

How do you bring in stakeholders at various points within the development process and ensure their different values and objectives are acknowledged, to efficiently deliver the outcomes? How do you ensure that a transparent WLV process is adopted throughout? This guide will have hopefully helped to achieve this.

'Initiative fatigue'

The advent of more and more initiatives and toolkits available to practitioners might lead to 'initiative fatigue'. This needs to be addressed by clients and practitioners when considering applying WLV principles on their projects.

Cost

Who pays and who benefits from WLV applications and how the benefits are shared? It might be useful to consider how the benefits can be shared between those who pay for WLV application from the outset of a project with those who benefit from such principles during operation and use.

Current budgetary arrangements

Budget constraint is in conflict with the WLV process – change may be required in the way budgets are prepared and viewed. When should budget be determined, especially for high-budget infrastructure development programme? For example, can capital and maintenance budgets be viewed as a single pot prior to completion of the project or is this unrealistic?

section 10
Case studies

Case study 1: A value-based selection methodology for products and services

Client: Home Group Ltd

Home Group Ltd sought to develop a way of identifying which products and services on key building components are likely to offer Home Group Ltd best value. This approach has been used by Home Group Ltd to enter into volume discount arrangements with selected manufacturers and suppliers. With the support of the Housing Corporation, the initial methodology has been developed into what can be a useful tool for other associations or specifiers.

The tool stems from work undertaken by Home Group Ltd (a group of major residential social landlords [RSLs – www.homegroup.org.uk], including Home Housing Association, Warden Housing Association, Stonham Housing Association and Home in Scotland) as part of its internal Egan Project, which was a major review of how Home Group Ltd could further respond to the Rethinking Construction agenda in its property development and maintenance activities.

The methodology and a spreadsheet tool are used to rank suppliers or components on the basis of the value that they offer, not just the prices that they charge. The methodology and tool focus on stage 1 of the procurement process, namely:

- Drawing up a long list of potential suppliers for each product area.
- Drafting value-based product selection criteria. These are criteria to assess product cost, product quality and service quality offered by the suppliers. The product and service quality criteria are mainly expressed as performance standards. The criteria are based on market and product knowledge of Home Group Ltd staff, ie maintenance staff, residents, in-house architects and some sub-contractor input.
- Inviting the long-listed suppliers to provide information about the product range of their choice, and applying the selection criteria to those products to arrive at a short-list of suppliers.

After this stage, short-listed suppliers are interviewed and invited to comment on the performance specification (stage 2) and firm prices are sought while KPIs are being developed (stage 3).

The assessment measures each potential supplier's scores against cost, quality and risk criteria, as indicated in Table 8. A default weighting for each criterion is identified, which can be changed by individual members of Home Group Ltd. Minimum acceptable scores can also be set for individual criteria.

After consultation within the maintenance teams, the working group identifies the following materials and components as those that accounted for the majority of maintenance expenditure in RSLs:

- Boilers
- Doors and windows
- Sanitaryware
- Decorative products (paints and stains)
- Kitchen units
- Electrical products.

A spreadsheet tool is used to complete scoring and assessment of each supplier against the agreed criteria and weightings. An example is shown in Table 9.

How these criteria related to individual products is agreed between stakeholders. An example of the specific criteria identified (for domestic boilers) is shown in Table 10.

Further information

- Home Group Ltd website: http://www.homegroup.org.uk
- Building Cost Information Service (BCIS) report to the Housing Corporation on WLCs available at http://www.housingcorplibrary.org.uk/housingcorp.nsf/AllDocuments/52474CEAE6829ADE80256DCC00530156/$FILE/Wholelife.pdf
- BCIS and the Building Maintenance Information (BMI) resource are available at http://www.bcis.co.uk.
- Of particular use might be the Life Expectancy of Building Components published by the BMI and available at http://www.bcis.co.uk/order/bmipub.html

Table 8 Selection criteria and weighting

Heading	Weighting
Cost	
● Capital cost	15%
● Maintenance, replacement and disposal costs	15%
● Costs in use	15%
Quality	
● Functionality	10%
● Frequency of renewal	5%
● Disruption to residents upon repair/renewal	10%
● Environmental sustainability	10%
● Aesthetics	10%
Risk	
● Proven track record/warranties	10%
Total	100%

Table 9 Specific selection criteria - boilers

Criterion	Assessment measures
Capital cost	● List price per boiler type (£) (prices for wall-mounted, floor mounted, back, combi and condensing boilers) ● Percentage reduction for a solus agreement ● Percentage reduction available as one of three named suppliers
Through life cost to landlord	● Servicing costs (1= good, 6 = bad) ● Failure rates (1= good, 6 = bad) ● Life expectancy (1= good, 6 = bad)
Through life cost to resident	● SEDBUK ratings (%) compared with the generic boiler type ratings ● Average percentage rating compared with generic type rating
Functionality	● Oxi-depletion device? ● Option for built-in and separate programmer? ● Electronic ignition?
Frequency of renewal	(Not separately assessed – scored in through life cost to landlord)
Disruption to residents	(Not separately assessed – assumed to be constant between all manufacturers)
Time taken to obtain the boiler	● Delivery time for appliance replacement ● Delivery time for replacement parts under guarantee
Replacement parts	● Delivery time for replacement parts not under guarantee ● Response time perceived by contractors (1 =good, 6 = bad)
Environmental sustainability	● Efficiency ratings (%) compared with the generic boiler type ratings ● Average percentage rating compared with generic type rating
Aesthetics	(Not assessed as all boilers perceived to be very similar visually)
Warranty	● Normal warranty period ● Availability of a national maintenance agreement?

Table 10 Example of assessment and weighting applied to alternative suppliers

Selection criteria				Supplier 1		Supplier 2		Supplier 3		Supplier 4		Supplier 5	
	Overall weighting	**Minimum threshold**	**Relative weighting**	**Score awarded (out of 10 for each)**	**Weighted score**	**Score awarded (out of 10 for each)**	**Weighted score**	**Score awarded (out of 10 for each)**	**Weighted score**	**Score awarded (out of 10 for each)**	**Weighted score**	**Score awarded (out of 10 for each)**	**Weighted score**
Cost criteria	40%	30%											
Capital cost for RS (initial provision and installation)			30%										
Capital cost assessment 1			**10**	7	70	6	60	6	60	4	40	10	100
Capital cost assessment 2			**10**	9	90	9	90	8	80	6	60	11	110
Capital cost assessment 3			**10**	6	60	9	90	5	50	4	40	12	120
Sub-total			OK		220		240		190		140		330
Maintenance, replacement and disposal costs (landlord costs)			45%										
Landlord revenue cost assessment 1			**20**	8	160	8	160	8	160	6	120	6	120
Landlord revenue cost assessment 2			**10**	9	90	9	90	9	90	7	70	7	70
Landlord revenue cost assessment 3			**15**	9	135	10	150	10	150	8	120	8	120
Sub-total			OK		385		400		400		310		310
Costs in use (resident costs)			25%										
Resident revenue cost assessment 1			**10**	8	80	9	90	7	70	8	80	8	80
Resident revenue cost assessment 2			**5**	5	25	10	50	8	40	9	45	9	45
Resident revenue cost assessment 3			**10**	9	90	11	110	9	90	10	100	10	100
Sub-total			OK		195		250		200		225		225

Case study 2: Value for money from the Highways Agency

Client: The Highways Agency

The Highways Agency is an executive agency of the Department for Transport and is responsible for the management of the strategic road network in England, a public asset worth over £65 billion (Figure 14). A key objective underpinning the Highways Agency's strategy for the management of the road network is given in the Highways Agency Business plan for 2004/05:

- To provide the tax-payer with value for money by maximising the efficiency of maintenance by keeping the network in a safe condition whilst minimising whole life costs, disruption to road users and others affected, and the adverse effect on the environment.

Vehicles contribute to the wear and tear on road pavements. This results in the need for maintenance and disruption to road users at the road works sites. Choices at the time of maintenance lead to different long-term impacts, on costs, road condition and service to road users. The Highways Agency has therefore adopted a WLC approach that takes account of both costs to the Highways Agency as well as costs to road users, to develop cost effective maintenance strategies for the pavement network.

The principal components of the WLC of road pavements are:

- Initial investment cost, ie the cost of construction for new pavements or the costs of the proposed maintenance option for an existing pavement.
- Future maintenance costs incurred during the life of the pavement.
- Road user costs ie costs incurred by road users at road works due to increased journey time and accidents.

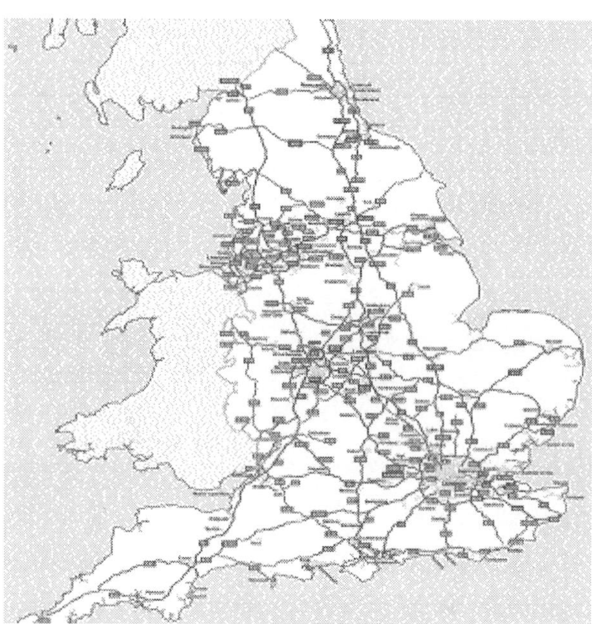

Figure 14 The strategic road network in England (Source: Highways Agency)

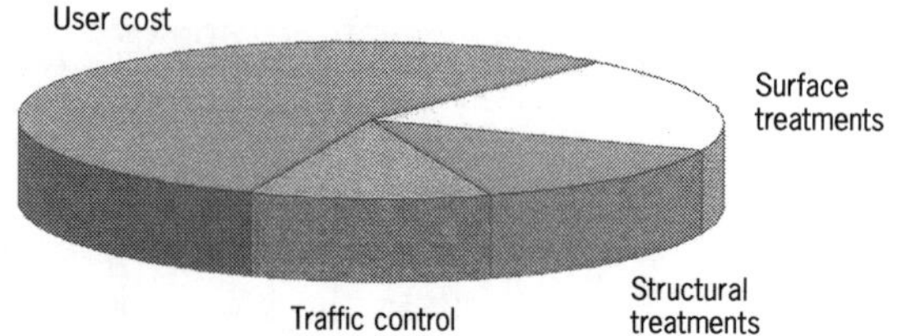

Figure 15 Road user costs on heavily trafficked roads

Meeting user requirements

The relative proportions of the components of WLC vary under different conditions of usage but road user costs can form a significant proportion of the total cost for highly trafficked roads (Figure 15). Disruption to road users can be minimised through alternative ways of working, eg carrying out maintenance at night when traffic flows are low, although these can result in higher works costs (Figure 16). The use of a WLC approach enables the identification of the option with the optimum combination of works and road user costs.

Value for money measure

In general, there is an economic benefit in carrying out the proposed maintenance only when the whole life cost of the proposed option is lower than that of the 'Do Nothing' option or 'Do Minimum' option, eg if pavement condition is below permitted functional levels set on the grounds of safety. An economic indicator (EI) representing the relative benefit of adopting the proposed maintenance option A rather than the 'Do Nothing'/'Do Minimum' (M) option is given by the reduction in WLC achieved by spending more initially:

$$EI = \frac{WLC_M - WLC_A}{\text{Initial cost of A} - \text{Initial cost of M}}$$

N.B. For option 'Do Minimum' (M) the initial cost is lower than 'Do Something' (A); but the WLCs of M might be higher than A, as in future it may require higher maintenance. Thus, the long life benefits of spending high (A) now are compared to benefits or savings of spending less now (M).

Figure 16 Choosing the time of carrying out maintenance to limit disruption to road users and works sites

Delivering value for money – at the strategic level
Performance targets and budget requirements are determined on the basis of minimum WLC taking account of costs to road users.

The Network Whole Life Cost Model is a prototype research tool, developed by the Highways Agency, which models the actual network (Figure 17). Data from the Highways Agency Pavement Management System, which describes the traffic levels, existing condition and rates of deterioration on each part of the road network, is used with the practical constraints that face highway engineers when developing the works programme, to identify the most cost-effective maintenance strategies.

Minimum WLC performance target
A new performance target for road pavement maintenance was introduced in the Highways Agency Business Plan for 2004 – 05. This was aligned with the key objective of maintaining the network at minimum WLC, ie minimising the combined direct cost of the works with the costs to the road user caused by those works.

The Network Whole Life Cost Model is used to determine the condition associated with the strategy that minimises the cumulative WLC of the maintenance programme. Targets are set typically for a period of 3 years using a staged analysis:

- Identification of potential maintenance lengths based on condition and defined standards, or condition thresholds, to meet safety and functionality requirements.
- For each maintenance option on each road length in the network, determination of the WLC of the following:
  - Maintenance option (including works and road user costs)
  - Associated 'Do Nothing'/'Do Minimum' option.
- Implementation of the option in each of the two alternative years of the 3-year programme period to identify the effects of delaying and/or bringing forward the works.
- Identification of the combination of the types of work and timing that gives the lowest cumulative WLC.

The predicted condition of the network represents the condition that minimises the WLC of the road network. Since 2000, the prototype Network Whole Life Cost Model has been used to support the Highways Agency bid for funding in the biennial spending review for the maintenance of the trunk road network at minimum whole life cost. In the future, the Network Whole Life Cost Model will be incorporated into the Highways Agency Pavement Management System.

Delivering value for money – at the programme level
Following the development of budgets at the strategic level, the maintenance programme is developed by a detailed examination of the maintenance options on each road length. The Highways Agency has developed a VM process to identify the maintenance that is expected to yield the maximum benefits to the road user giving best value for money and is consistent with the Highways Agency strategic aims.

The VM process (Table 11) is used to carry out relative assessments of all road maintenance bids. Each bid is scored against six criteria that are weighted according to their relative importance:

- Safety (0.3)
- Value for money (0.2)
- Level of service (0.2)
- Environment (0.1)
- Network disruption (0.1)
- Quality of submission (0.1).

The resulting score is used to prioritise the Highways Agency Road Maintenance Programme across the network. The value for money score is derived from analyses of the WLC of the maintenance options, including the 'Do Nothing'/'Do Minimum' option, on each road length, using the specially developed Scheme Analysis System (SAS). At the start of 2005, this stand-alone spreadsheet based system will be replaced by a more detailed version of SAS that will be available to engineers as part of the Highways Agency Pavement Management System.

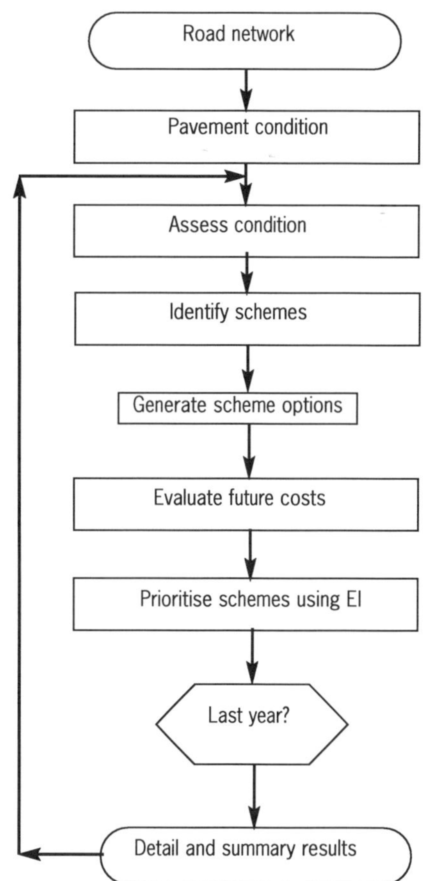

Figure 17 The Network Whole Life Cost Model

Table 11 VM scoring example

Safety	There is no specific accident problem at this site. Skid resistance levels are all above the investigatory level for this type of non-event carriageway and the texture of the existing hot-rolled asphalt surfacing is reasonably good. As the site has a lower than average level of accidents for the type of site, it has a poor safety justification. **Score: 30**
Value for money	The cost of the works compared to the value of the pavement is low. The whole life cost analysis of different treatment options indicates that the proposed type and timing of maintenance offers the best value for money. Value for money is increased by undertaking maintenance of the vehicle restraint system at the same time as the carriageway works. **Score: 70**
Level of service	The (13-year-old) HRA surfacing is starting to deteriorate but there have to date only been limited emergency works which have all been at night and have caused only limited disruption. This is not expected to change significantly over the next year. Therefore, the works are likely to have little or no effect on the current level of service. **Score: 30**
Environment	The inlay will use a low noise surfacing. However, the road is in a rural area, there are limited numbers of dwellings likely to be affected. No evidence is provided that the inlay can make use of recycled materials. There is therefore only a small positive environmental effect. **Score: 40**
Network disruption	There will be considerable disruption caused by this maintenance. However, there is little or no scope for reducing this by using other arrangements (e.g. local residents prevent night working and off-peak only working is prevented by the traffic flows). **Score: 80**
Quality of submission	The information used to determine the maintenance solution included a full set of confirm data and further detailed surveys. The information was well presented and strongly supported the proposed maintenance solution (ie inlays to remove cracked and rutted material). **Score: 75**

	Safety	**Value for money**	**Level of service**	**Environment**	**Network disruption**	**Quality of submission**	**Total**
Score	30	70	30	40	70	75	
Weighting	0.3	0.2	0.2	0.1	0.1	0.1	
Weighted score	9.0	14.0	6.0	4.0	7.0	7.5	47.5

Case study 3: The A303 Stonehenge Improvement

Client: Department for Transport

The Stonehenge Project (formerly the Stonehenge Master Plan) comprises three elements: the A303 Stonehenge Improvement, the new visitor centre and the National Trust Land Use Plan. Case study 3 is concerned with the A303 Stonehenge Improvement. WLV principles have been used in option appraisals for this scheme from conception to detailed engineering design. The A303 Stonehenge Improvement is being promoted by the Highways Agency as an 'exceptional environmental scheme' in recognition of its unique context within the Stonehenge World Heritage Site (WHS).

The implementation of the Stonehenge Project is overseen by representatives from organisations including: Council for British Archaeology; English Nature; Environment Agency; Highways Agency; Ministry of Defence; National Trust; Prehistoric Society; Salisbury District Council; Wiltshire County Council; residents of Amesbury (the nearest town); road users; and the general public.

The visual, environmental, and safety impacts of existing road traffic from the existing A303 and A344 on Stonehenge and associated monuments would be mitigated by means of a 2.1 km tunnel which would also allow the unique historical landscape to be re-created. Stakeholders including the National Trust, English Heritage, the Environment Agency and English Nature have been closely involved with the scheme's development. Their inputs have enabled the Highways Agency to promote a scheme which best balances all the potential benefits at an affordable WLC. The scheme would bring benefits to the Stonehenge landscape and the village of Winterbourne Stoke.

The A303 trunk road is an important route between the M3 and the West Country, and is part of the Trans European Route Network. The 9 km (5.5 mile) section west of Countess roundabout is the first section of single carriageway encountered by westbound traffic and upgrade of the current situation is required in order to address the following issues:

- An average of 32 000 vehicles per day causing traffic congestion, noise and air pollution.
- The current road capacity is insufficient for predicted increases in demand.
- The route currently passes through Winterbourne Stoke, a rural community, concerned about noise, pollution, and safety.
- The accident record of the existing road is poor.
- Road traffic causes visual and noise intrusion at Stonehenge and divides it from surrounding associated monuments in the WHS.

The Stonehenge Project aims to remove 20th century 'clutter' from sight of Stonehenge and to enable recreation of the historical landscape. The objective of the Stonehenge Project is that "measures should be identified which will provide comprehensive treatment of important road links within the WHS in order to reduce traffic movements and congestion, improve safety and enhance the historic environment." (Highways Agency)

The scheme entails the following:

- Converting 12.4 km into dual carriageway, increasing capacity.
- Building 2.1 km of tunnel near Stonehenge to take the road out of sight.
- Applying archaeological mitigation strategy and agreed soil handling strategy to provide effective mitigation of effects.

The Highways Agency has adopted a partnering approach with the major stakeholders. Consultation has taken place and has been organised with participation from national to local community members. The draft orders and environmental statement were published in June 2003 and this was followed by a public inquiry in 2004. Comments from all stakeholders were taken into account in developing the published scheme.

Stage 1: Inception

Throughout the 1990s various proposals were considered. Two were short-listed. Long-term impacts on heritage and traffic were considered during initial design. The challenge throughout the project has been balancing the potential environmental impacts and particularly cultural heritage requirements with affordable construction. After much debate and consideration, the Stonehenge Project (then known as the Master Plan) was announced by the government in 1998 and incorporated a 2 km cut and cover tunnel, which was at that stage supported by both English Heritage and the National Trust.

Stage 2: Option appraisal

Further consultation took place with the general public and with government and public bodies. Various configurations of tunnel length, junctions and roundabout locations were considered. Following representations from all consultees, the Published Scheme includes a 2.1 km bored tunnel, which is supported by English Heritage but opposed by the National Trust, who argued in favour of a longer tunnel.

Stage 3: Design development

WLC techniques have been applied to analyse the long-term costs and savings to the Highways Agency and road users. The 'Do Minimum' option was evaluated based on the lifelong impacts and savings of not carrying out any construction, compared to the lifelong impacts and benefits of the proposed 2.1 km tunnel. The Published Scheme, incorporating a 2.1 km tunnel, was tested under several weeks of cross examination at the recent public inquiry. It will be for the Inspector and Secretary of State to determine, based on all of the evidence presented at the inquiry, whether the Published Scheme with its 2.1 km tunnel provides best value.

Proposals to extend the tunnel length to 2.67 km or 4.5 km were put forward by the National Trust and developed by the Highways Agency. The costs and benefits of each tunnel variant were calculated using WLC methodology and COBA guidelines.*

The environmental impacts of each variant were assessed and compared using the methodologies set out in Volume 10 of the DMRB and a multiple criterion appraisal was presented comparing permanent and construction impacts. In this analysis the options were compared using each of the standard DMRB topic headings including cultural heritage, landscape, water, noise, air quality, biodiversity, policies and plans and the effect on vehicle travellers.

VM has been incorporated throughout the project development by considering qualitative and quantitative value drivers based on GOMMS (*Guidance on Methodology Multi-modal Studies*). GOMMS was used to derive an Appraisal Summary Table (AST) in which the Published Scheme was compared against the 'Do Minimum' across a range of non-financial quantitative measures of environmental impact.

In terms of RM, costs and risk allocation of all options were identified using COBA and risk assessment techniques. The resultant benefit to cost ratio developed identified the 2.1 km tunnel as having the highest and best ratio. The options for tunnel construction technique also focused on mitigating impacts associated with bored tunnel versus cut and cover. The results of COBA provide one measure of the NPV of the Scheme taking into account its capital cost compared to the user delay, vehicle operating and other benefits that flow from it. Because the A303 Stonehenge Improvement is a Special Environmental Scheme, in which around one third of the funding would be from heritage sources, the NPV and Benefit Cost Ratio (BCR) when assessing the Published Scheme as a transportation investment are, as expected, less than would apply to most schemes in the Highways Agency's programme. Nevertheless, the Published Scheme shows a positive BCR for high traffic growth and its BCR was better than any other deliverable alternative presented to the Inquiry.

Stage 4: Construction

Continuing with WLV analyses, to maintain user benefit and minimise WLCs, a service tunnel and a fire suppression system have been proposed. The plan incorporated WLCs and benefits of the two systems over the expected life of 120 years. The service tunnel will facilitate maintenance of the trunk road tunnel, greatly reducing user delays and maintenance costs. The fire suppression system has an intrinsic safety value throughout the life and use of the tunnel.

Service tunnel

The breakdown of tunnel maintenance lifetime costs by activity is shown below. The high proportion of renewal costs (Figure 18) emphasises the benefit of long-life, efficient systems, that can be easily and regularly maintained (Table 12).

Table 12

Total additional capital cost for service tunnel compared to direct road tunnel maintenance	£1 803 000
Savings due to avoiding direct road tunnel maintenance costs, closures and delays throughout life	£6 726 170
Thus, the service tunnel offers a **saving** of	£4 923 170

Source: From working paper by Halcrow – *Benefits of a Service Tunnel*

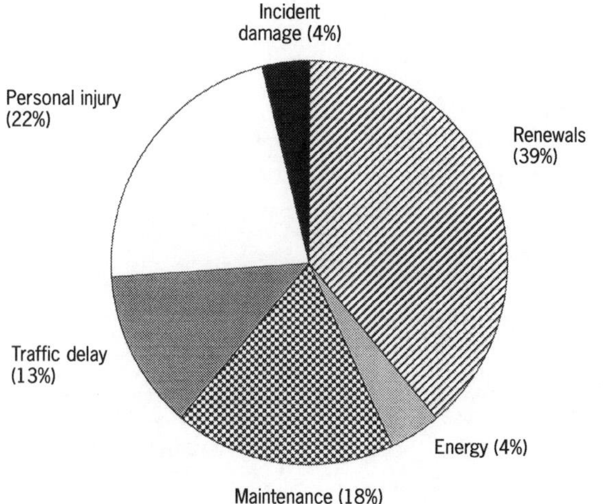

Figure 18 Share of costs by activity (Halcrow: *Benefits of a Service Tunnel*)

* (Trunk road schemes are routinely appraised by a specific cost-benefit procedure designed by the Department for Transport, known as COBA. The COBA methodology includes an 'indirect tax correction factor' which is used to scale down households' valuations of non-working time: www.dft.gov.uk/stellent/groups/dft_econappr/documents/page/dft_econappr_504872.pdf.)

Fire suppression system

WLC techniques are being used to determine whether a fire suppression system should be installed, and if so, what type. Preliminary results suggest that manually operated fire suppression would provide a positive return.

Breakdown of incident benefits through cost reduction from different cost sources are detailed in Table 13 and Figure 19.

Table 13

The average cost of construction	£6 300 000
The average benefits in reduction of injury and delay	£9 200 000
Thus, the fire suppression gives a **benefit** of	£2 900 000

Source: From working paper by Halcrow – *Benefits of a Service Tunnel*

Thus, the A303 Stonehenge Improvement has adopted WLV principles from conception right through to decision making about support systems needed to construct long life sustainable and reliable tunnels.

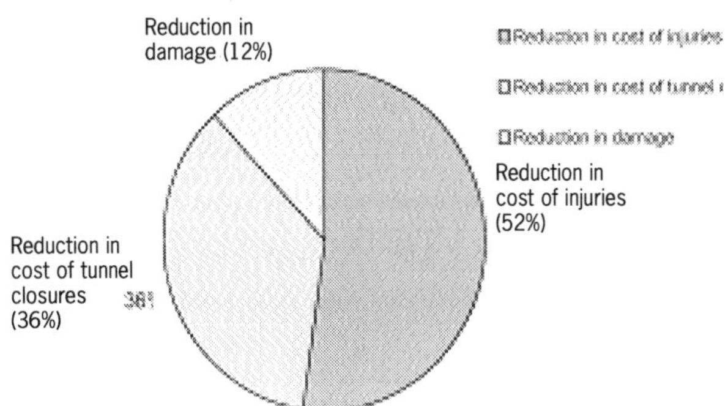

Figure 19 Reduction in costs (Halcrow: *Benefits of a Service Tunnel*)

Case study 4:Whole Life Value profiling: pilot new schools PFI project

Client (confidential)

This case study has been provided by representatives of a Whole Life Value Task Force set up by the European Construction Institute (ECI). The Task Force developed the concept of integrated WLV Profiling (WLVP), which used recognised balance scorecard and performance benchmarking techniques. A resulting WLVP approach was developed by the ECI Task Force – in association with Faithful & Gould (who invented the concept – www.fgould.com), with support from Constructing Excellence, RICS and CIC Design Quality Indicators and other major players in the construction industry.

The approach is a simple yet effective way of presenting all of the factors affecting value, in a form that is easy to understand and can show the impact of satisfying stakeholders' conflicting interests, by comparative WLV profiles. The resulting tool developed enhances decision making, and helps visualise options evaluations, assisting stakeholders to make informed choices.

The WLVP approach considers the multitude of project stakeholders, who can have differing interests and influences over the decision-making process. The WLVP tool harnesses the capability that already exists in the industry, ie using Design Quality Indicators, to provide the opportunity to bridge the communication barriers between customers and providers in the process.

The project aim was to build five new build schools – two secondary and three primary schools under the PFI procurement method. The WLV parameters were defined by the project sponsor in their outline business case and evolved into more detailed specific requirements in the PFI procurement documentation.

For example for a project in the North of England, base date 2003:

- Ready for use in the third quarter of 2007
- Required usage life of 25 years concession
- Construction period of 24 months.
- Capital cost, outline business case with a target of £50m capital expenditure.

Costs

- Cost in use of £1 million per annum
- £7million unitary charge per annum.

Compliance

- Deadline of third quarter of 2007 or miss the start of New Year.
- Environmental rating, excellent.
- Mandatory space standards, within the DFES guidelines of BB98/99
- Hand back obligations plus a 5-year life remaining in the asset.
- Community integration and third party use out of school hours.

One of bidders used the WLV principles, firstly to shortlist the conceptual design proposals in consultation with the school representatives and key project stakeholders. The shortlisted schemes were then priced and compared to the project sponsors declared affordability limits – which identified and scored the whole life costs versus the set cost parameters.

All the preferred schemes were then scored, using the WLV toolkit, and the outputs compared to the target parameters in the form of WLVP. The functionality, impact and build quality was scored using the CIC design quality indicators tool, which is a mandatory requirement for all DFES projects.

The aggregate score was used to establish the WLV score for each of these value drivers, for each option (Figure 20). Various more detailed options were then run to refine and optimise the preferred schemes – which identified what the client could afford, along with specific trade-offs, or shopping list client choices (Figure 21 and 22). For example: less area – changing the building shape, height, site position; specification choice – extent of fixtures and loose equipment, extent of external landscaping and play areas.

Where the scores are not identical it indicates a difference between the client requirements and the options score for that criteria (see Figure 20, 21 and 22).

Key benefits

The use of WLV approach made a unique difference in various ways, such as:

- Stakeholder involvement: understanding their needs versus wants and option selection.
- Setting WLV principles to identify the preferred solutions and specific choices.
- Achieve the best value solution within the set parameters and affordability limits.
- Improve the effectiveness of the PFI evaluation and investment decision making process.

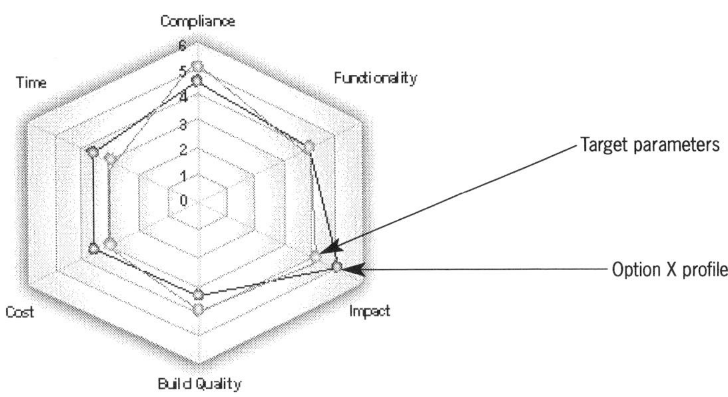

Figure 20 The factors affecting value

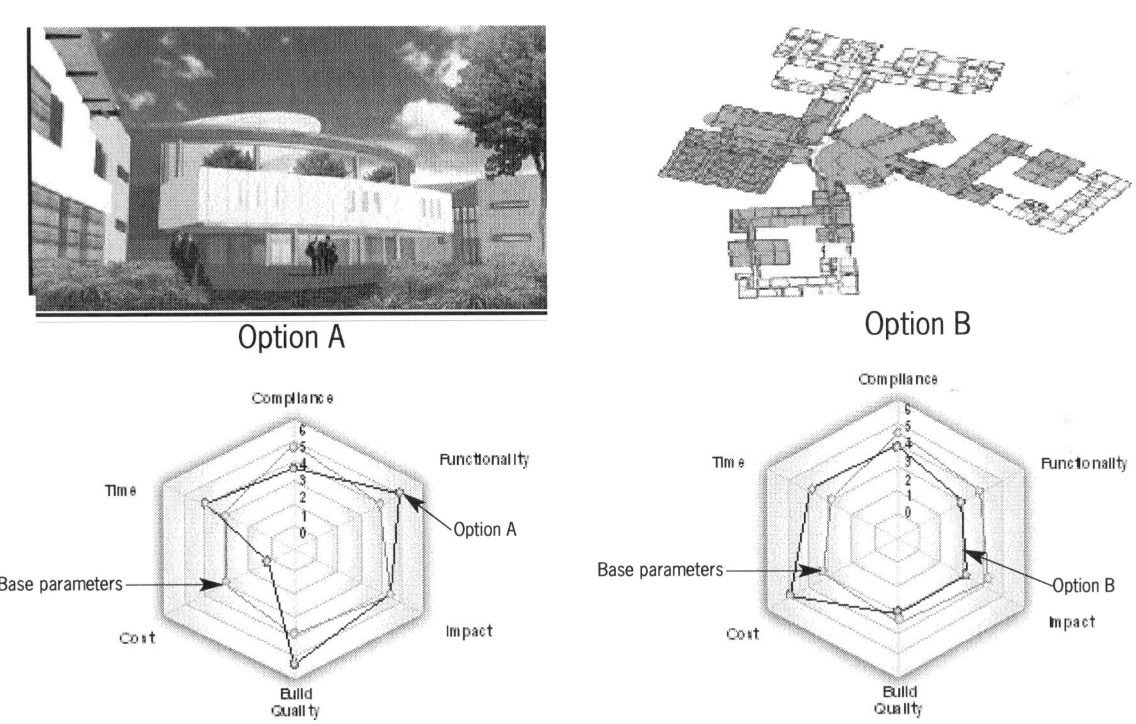

Figure 21 Comparison of base parameters versus new build

Figure 22 Comparison of base parameters versus new build

annex 1
Key WLV drivers in the public sector

The following organisations, policy initiatives and rules have had a significant and ongoing effect on public sector procurement, and help to drive the adoption of WLV as the basis of public sector procurement. This directly also affects the private sector as partners and suppliers.

Organisations

HM Treasury

The Treasury is the UK's economics and finance ministry. It is responsible for formulating and implementing the government's financial and economic policy. Its aim is to raise the rate of sustainable growth, and achieve rising prosperity and a better quality of life with economic and employment opportunities for all.

The *Treasury Green Book - Appraisal and Evaluation in Central Government* (HM Treasury, 2003) is a key document determining how value for money in public sector investment is appraised, and also contains guidance on discounting, risk, optimism bias etc.

National Audit Office

The National Audit Office (NAO) is an independent public body responsible for ensuring that public money is spent economically, efficiently, and effectively in the areas of local government, housing, health, criminal justice and fire and rescue services.

The role of the NAO is to audit the financial statements of all government departments and agencies, and many other public bodies. It also reports to parliament on the efficiency with which these bodies have spent public money. As well as providing accountability to parliament, it aims to bring about real improvements in the delivery of public services.

The Auditor General for Wales

The primary role of the Auditor General for Wales is to provide independent information, assurance and advice to the National Assembly on the way in which it, and other public bodies in Wales, account for and use taxpayers' money. The Auditor General, supported by the National Audit Office, also aims to help such bodies to provide better financial management and value for money.

Audit Scotland

Around 200 public bodies in Scotland spend over £20 billion of public money each year. Examples include the Scottish Executive, local councils and health boards. The duty of the Auditor General and the Accounts Commission is to check that public money is spent properly, efficiently and effectively. Audit Scotland's role is to provide the Auditor General and the Accounts Commission with the services they need to carry out their duties.

Northern Ireland Audit Office

The Northern Ireland Audit Office's objectives are to provide effective support to the Northern Ireland Assembly in its task of holding departments and agencies to account for their use of public money and effective local government audit. It also aims to provide support to the Northern Ireland public sector bodies in their pursuit of improved financial reporting and value for money, including support for efforts to combat public sector fraud.

Office of Government Commerce

The Office of Government Commerce (OGC) is an independent office of the Treasury reporting to the Chief Secretary. It is responsible for a wide-ranging programme, which focuses on improving the efficiency and effectiveness of central civil government procurement. In addition, the OGC has an important role in developing and promoting private sector involvement across the public sector.

Since 2003–04, the OGC has also assumed a key role in assisting departments set up project and programme management centres of excellence in their departments. These new programme management centres will become central points for embedding project and programme management best practice across government.

4ps

4ps is the local government procurement expert, providing advice, guidance and skills development to local authorities undertaking projects, procurement and partnerships. This includes the PFI schemes, strategic service partnerships and all other forms of partnership working. With extensive experience and expertise in developing, procuring and delivering large, high risk, complex projects, 4ps offers comprehensive procurement support to local authorities, including hands-on project support, Gateway™ Reviews, skills development and 'know-how' procurement guidance in the form of procurement packs, case-studies and Extranets.

Policy initiatives

Achieving Excellence in Construction

Government departments have a wide range of possible options when purchasing. Outsourcing and the PFI have shifted the focus to thinking in terms of whole services such as the management of prisons rather than buying the goods and services necessary to deliver or manage the service. HM Treasury recommends a comprehensive procurement strategy to cover this new holistic approach that should include:

- an analysis of key goods and services, their cost and priorities, which the department or agency needs to deliver its objectives and services.
- an assessment of the way these are purchased.
- the performance of key suppliers.
- the scope to improve value for money and quality of service.

The Chief Secretary to the Treasury launched the initiative, Achieving Excellence, in March 1999 to improve the performance of central government departments, executive agencies and non-departmental public bodies as clients of the construction industry (*Property and Construction Best Practice Development Centre – Achieving Excellence in Construction*, OGC, 1999).

The publication *Achieving Excellence in Construction* is associated with a suite of procurement guidance, which aligns with the Gateway™ Process and Successful Delivery Toolkit developed by the OGC.

Achieving sustainability in construction procurement

In the move towards sustainable construction, construction industry clients have a key role (Government Construction Clients' Panel [GCCP], 2000). Responses to the government's Opportunities for Change consultation stressed that the industry expects government to take a lead as clients for public works, as regulators and legislators. The GCCP Action Plan underlined this commitment by setting up a sustainability action group to investigate how construction procurement can contribute to policy in sustainability.

Although the overarching aim of procurement is more focused on the achievement of value for money than on delivery of policies such as environmental sustainability, there is much scope to consider sustainability issues within the value for money approach.

As with Achieving Excellence, construction procurement is taken to include new works, refurbishment and maintenance projects. To move and to measure progress in a sustainable direction, a framework and a set of goals are needed. The framework used is based upon 10 themes for action included in the strategy for more sustainable construction Building a Better Quality of Life (DETR, 2000).

Included are suggested commitments that will result in:

- procurement in line with value for money principles on the basis of whole life costs.
- less waste during construction and in operation.
- targets for energy and water consumption for new projects that meet at least current best practice for construction type and which contribute significantly to the achievement of cross-government targets.
- the protection of habitat and species taking due account of the UK Biodiversity Action Plan and the biodiversity action checklist for departments.
- targets developed in terms of 'respect for people' for procurement of the government estate.
- a contribution to the goals of less pollution, better environmental management, and improved health and safety on construction sites.

The consultation to develop a new UK sustainable development strategy was launched on 21 April 2004 and responses have been taken on board for the new UK government strategy, which will be launched in Spring 2005.

Best value in local government

From 1 April 2000 the government placed a new duty of best value on local authorities, establishing challenging new arrangements under which they will fund, procure and deliver all of their services (Institution of Highways and Transportation, 2001). It requires local authorities to:

- ensure that services are responsive to the needs of citizens, rather than the convenience of service providers.
- secure continuous improvement in the exercise of all functions undertaken by the authority, whether statutory or not, considering a combination of economy, efficiency and effectiveness.

Other principles of best value include:

- ensuring that public services are efficient and high quality.
- ensuring that policy making is more joined up and strategic, forward looking and not reactive to short term pressures.
- using information technology to tailor services to the needs of users.
- valuing public services and tackling the under representation of minority groups.

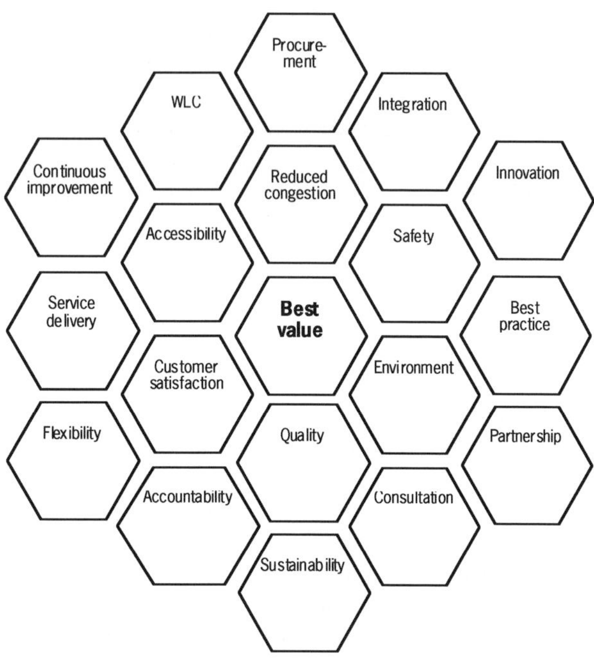

Figure 23 The best value framework (source: Institution of Highways and Transportation, 2001)

Although best value applies only to services provided by local authorities, the Highways Agency in England and the strategic roads authorities within the devolved administrations are applying similar principles and priorities. The Highways Agency strategy for investment is also strongly focussed on the requirements of users and based on:

- making road maintenance the first priority.
- making better use of the roads, through network control, traffic management measures and safety improvements.
- WLC, seeking to minimise costs over time taking into account the effects of disruption to traffic.
- tackling some of the most serious and pressing problems through a carefully targeted programme of small improvements.

The aim of the Best Value Programme run by the Office of the Deputy Prime Minister (ODPM) for local government, is to improve delivery and value for money of local services through implementing a Comprehensive Performance Assessment, engaging with authorities in improvement planning, negotiating and monitoring local performance agreements, providing a package of freedoms and flexibilities, developing capacity building programmes and supporting the electronic delivery of services.

Best value indicators are measures of performance set by the departments in local government (Figure 23). They are called Best Value Performance Indicators or BVPIs. They have been set since the duty of best value on local authorities came into effect under the Local Government Act 1999. (For further information see the ODPM Local Government Performance site www.bvpi.gov.uk.)

Guidance on how to meet the best value requirements of the Local Government Act 1999 is available from the ODPM, *Circular 03/2003* (ODPM, 2003). The guidance replaces earlier circulars published by the DETR, and applies to all principal local authorities in England.

Best value in England

- Statutory basis Local Government Act 1999
- Best Value Performance Plans
- Reviews of all services on a 5-year cycle
- Statutory inspection by Audit Commission
- Statutory framework of BVPI.

Best value in Wales

- Statutory basis Local Government Act 1999
- Best Value Performance Plans
- Reviews of all services on a 5-year cycle
- Audit Commission have role but no statutory requirement for inspection
- Statutory framework of BVPIs.

Best value in Scotland

- No statutory basis at present but legislation is planned. The moratorium on CCT is continuing on the basis that all authorities have adopted the best value regime.
- No requirement to publish best value performance plans. Authorities are required to have in place public performance reporting frameworks, which are subject to audit.
- No statutory requirement for best value reviews, but authorities have agreed with the Scottish Executive to review services on 4 to 5 year cycle.
- Audit Scotland has a role but no statutory requirement for inspection.
- The Local Government Act 1999 lays down statutory performance indicators, and Public Performance Reporting Framework includes for additional key performance indicators, but little relevant at present to highway maintenance.

Review of civil procurement in central government

About £13 billion per year is spent by the UK on civil procurement (that is excluding military equipment), which is the equivalent of 2 to 3 pence in the pound on income tax. Of this, about £7.5 billion is spent on construction. Procurement covers every aspect of the process of determining the need for goods and services, and buying, delivering and storing them.

Procurement is defined as the whole process of acquisition from third parties (including logistical aspects) and covers goods, services, and construction products. This process spans the whole life-cycle from concept and definition of business needs, through to the end of the useful life of an asset or the end of a services contract (Gershon, 1999).

Modernising construction

In 2001 the NAO recommended that procurement processes concentrate on the following (*Modernising Construction*, NAO, 2001):

- The basis of value for money instead of lowest price.
- Closer working between clients and all those involved in the design and construction.
- Integration of the supply chain to drive out waste and reduce the costs of the construction throughout its whole life.

Value for money in procurement is not about achieving the lowest initial price, rather it is defined as the optimum combination of whole life costs and quality.

Review of public sector efficiency 2003 – ongoing

In August 2003 the Prime Minister and Chancellor commissioned an independent efficiency review looking across central, regional and local government, and the wider public sector, to identify savings and efficiency gains that would release essential resources to front-line services. The remit was to scope for efficiency savings across all public spending within departmental spending limits, and the results were to feed directly into the Chancellor's 2004 spending review report. The efficiency review team produced robust and detailed proposals that recommended to ministers stretching but realistic departmental efficiency targets for the period 2005–06 to 2007–08 which will deliver gains of over £20 billion. For more information see www.ogc.gov.uk.

Rules

European Union (EU) procurement directives

Opening public procurement to competition from all the member states is an integral part of the European Single Market. There are separate EU procurement directives and regulations which implement them in the UK, covering supplies, works and services contracts in the public sector. In addition there is a directive and regulation covering the utilities sector. These directives, and their associated regulations, require that contracts above specified thresholds are advertised in the *Official Journal of the European Communities* and they include detailed requirements for the selection of bidders and the award of contracts, based on the principles of transparency, non-discrimination and competitive procurement (NAO/OGC, 2001).

The existing EU directives covering procurement of services, supplies and works are to be replaced by a new consolidated public sector directive. The new directive will for the first time include a provision on framework agreement contracts for the first time. For more guidance on EU procurement rules, see the OGC publication *Tendering for Government Contracts – a Guide for Small Businesses*, available via www.ogc.gov.uk, which includes many useful references.

The directives allow purchasers to award contracts on the basis of either price or the economically most advantageous tender. The latter is equivalent to value for money, or best value and it should be used by UK contracting authorities.

HM Treasury procurement rules

HM Treasury *Procurement Guidance Note No.7 on Whole Life Costs* (November, 2000) now demands that "all (public sector) procurement must be made solely on the basis of value for money, in terms of Whole Life Costs". This means that there is a duty on suppliers to deliver services to clear standards that cover both cost and quality by the most effective, most efficient, and most economic means available. While not explicitly stated, this also embraces sustainability issues.

The Gateway™ Process

The OGC's Supervisory Board, chaired by the Chief Secretary to the Treasury, agreed that from January 2001 the Gateway™ Process would be mandatory for all new high-risk projects. In fact, from mid-2001, all medium risk projects and from early 2002, all low risk projects which involve procurement in the public sector would be reviewed by the process (NAO/OGC, 2001).

In simple terms, the OGC Gateway™ Process (www.ogc.gov.uk/sdtoolkit/workbooks/gateway/index.html) is a review of a project carried out at a key decision points or gates by a team of experienced people independent of the project team on behalf of the project sponsor. The purpose is to ensure that the project is justified and that the proposed procurement approach is likely to achieve value for money.

annex 2
List of tools

How to use the list of tools

- Use the name of the tool or tick list columns to determine which tools are relevant to your enquiry. You can use the following criteria:
  - Tool name
  - Who will use the tool (designer, clients, contractors, others)
  - When the tool will be used in the project (inception, feasibility appraisals, plan and design, operate and maintain, construct and handover, decommission and renewal
  - What type of projects will it be used for (buildings and/or infrastructure)
  - Which key processes are relevant (WLC, LCA, MCA).
- Check the brief description.
- If the description appears relevant to your enquiry check the reference column for a website link or means to find out more.

Tool	Who will use this tool?				When will it be used?						Project types?		Relevance to WLV?		
	Designer	Client	Contractor	Other	Inception	Feasibility appraisals	Plan/design	Construct/handover	Operate/maintain	Decommission/renew	Buildings	Infrastructure	WLC	LCA	MCA
A client's guide to greener construction *www.ciria.org.uk* This guide covers key issues from legal requirements, transport and local environment to conserving energy and water, dealing with waste and selecting appropriate materials. The guidance is supported by illustrative case studies, which demonstrate the environmental and potential economic benefits available from a positive environmental approach.		✓			✓	✓			✓	✓	✓			✓	✓
A guide to procuring local authority transport schemes and services *www.4ps.gov.uk/Home.aspx?PageID=4.3.2&ctl=PubsDetails&PubsID=108* This guide on procurement options seeks to assist Local Authorities in understanding the range of options available to deliver services in the most efficient manner possible.		✓				✓						✓			✓
BREEAM (Building Research Establishment Environmental Assessment Method) *http://products.bre.co.uk/breeam/index.html* BREEAM provides authoritative guidance on minimising adverse effects of buildings on the local and global environments. The assessment is based on 'credits' awarded for a set of performance criteria. The outcome is a certificate or label that enables owners or occupants to gain recognition for their building's environmental performance. Based on the building design, BREEAM assesses the environmental performance of buildings on a scale from fail to excellent. There are versions for existing and new offices, dwellings (EcoHomes), superstores and retail (under development).	✓	✓	✓	✓	✓	✓	✓				✓			✓	
BS ISO 15686-1 Buildings and Constructed Assets – Service Life Planning – General Principles *BS/ISO 15686-1 www.bsi-global.com* International standard on service life planning - including how to design and select components and assemblies for their required design life, and how to estimate service lives using factors.	✓	✓	✓	✓		✓	✓				✓	✓	✓	✓	
BS ISO 15686-5 Buildings and Constructed Assets – Service Life Planning – Whole Life Costing *BS/ISO 15686-5 www.bsi-global.com* BSI Handbook of draft International standard on service life planning - whole life costing. Including equations, measures and examples, and guidance on how to use whole life costing as part of planning the service life of a facility.	✓	✓	✓	✓		✓	✓				✓	✓	✓		
BEES (Building for Environmental and Economic Sustainability) *www.bfrl.nist.gov/oae/software/bees.html* BEES (version 2:0) is an interactive computer design aid that helps users select building products for use in commercial office and housing projects in a way that balances environmental and economic criteria.	✓			✓		✓	✓				✓			✓	

Tool	Who will use this tool?				When will it be used?						Project types?		Relevance to WLV?		
	Designer	Client	Contractor	Other	Inception	Feasibility appraisals	Plan/design	Construct/handover	Operate/maintain	Decommission/renew	Buildings	Infrastructure	WLC	LCA	MCA
CEEQUAL: A civil engineering environmental quality assessment and award scheme *www.ceequal.com* CEEQUAL is an awards scheme assessing the environmental quality of civil engineering projects – a civil engineering equivalent to BREEAM. It is promoted by ICE, BRE, CIRIA and other industry organisations. It encourages environmental excellence, and improved environmental performance in project specification, design and construction. CEEQUAL uses a credit-based assessment framework, applicable to any civil engineering project, and includes environmental aspects such as use of water, energy and land as well as ecology, landscape, nuisance to neighbours, archaeology, waste minimisation and management, and amenity. A CEEQUAL award recognises the achievement of environmental performance. Awards are made to projects in which the clients, designers and contractors go beyond the legal and environmental minima to achieve distinctive environmental standards.	✓	✓	✓		✓	✓	✓	✓	✓	✓		✓		✓	✓
Clients guide to functionality *BRE Report 452 www.brebookshop.com* This guide shows how the 'functionality' of construction products can add value, so it can become an intrinsic part of both the decision to build and the briefing process. The guide shows clients and suppliers the importance of functionality in 'construction' – all built facilities, including buildings and infrastructure. The guide includes details of steps which can be taken and tools available to enable functionality to be considered as part of the construction briefing process, and of the business review process.	✓	✓			✓	✓					✓	✓			✓
Code of Practice on the Dissemination of Information *www.odpm.gov.uk/stellent/groups/odpm_planning/documents/page/odpm_plan_606210-01.hcsp* This code of practice addresses the impacts of construction projects on people's daily lives and sets out a process for reducing their effects through communication and information provision. It sets out suitable actions to be undertaken at each stage of a development to keep the public informed and involved.	✓	✓	✓	✓		✓	✓	✓		✓		✓			✓
Community Impact Evaluation (CIE) *Community Impact Evaluation (CIE). Tecla Mambelli, 30 May 2000* CIE takes account of the total costs and benefits on a community and brings out the incidence of these on the various community sectors. The analysis identifies the sectors (producers on-site and off-site and consumers) that would be affected by a project or a plan, describes the kind of impact on them, defines their sectoral objectives, notes the unit of measurement and valuation of the impact and indicates the sectoral preferences on project alternatives.		✓		✓	✓	✓	✓	✓			✓	✓			✓

Tool	Who will use this tool?				When will it be used?						Project types?		Relevance to WLV?		
	Designer	Client	Contractor	Other	Inception	Feasibility appraisals	Plan/design	Construct/handover	Operate/maintain	Decommission/renew	Buildings	Infrastructure	WLC	LCA	MCA
Construction Procurement Guidance – No 7 Whole Life Costs *www.ogc.gov.uk/sdtoolkit/reference/achieving/ae7.pdf* This guidance sets out the principles for preparing whole life cost models. The former central unit on Procurement (now part of OGC) Guidance Note No 35 Life Cycle Costing provides examples of WLC though they are not related to construction projects. This is the seventh in a series of ten documents that translate the recommendations in the Efficiency Unit Report *Construction Procurement by Government* into practical proposals. It addresses the recommendations made in the University of Bath School of Management Agile Construction Initiative report *Constructing the Government Client* and in the Construction Taskforce report *Rethinking Construction*.		✓		✓	✓	✓	✓	✓			✓	✓	✓		
Dashboard of Sustainability *www.iisd.org/cgsdi/dashboard.asp* The Dashboard of Sustainability is a free software tool, which allows the presentation of complex relationships between economic, social and environmental issues aimed at decision-makers interested in sustainable development. Using the metaphor of a vehicle's instrument panel, it displays country-specific assessments of economic, environmental, social and institutional performance toward (or away from) sustainability.	✓	✓		✓	✓	✓	✓			✓	✓	✓	✓	✓	✓
Design Quality Indicators, Construction Industry Council *www.dqi.org.uk* Online design quality indicators to improve the quality and value of buildings. DQI Online is an interactive tool with a simple, non-technical questionnaire. The process of answering the questions helps an assessment of the quality of a building to be made in an interactive and participative way, so all stakeholders can get involved. The results can be obtained instantly and displayed in different ways to help facilitate discussion. The questionnaire addresses three critical areas: functionality, build quality, impact. It also encompasses the wider effect its design may have within the design and construction community.	✓	✓	✓	✓			✓				✓				✓
Ecopoint *BRE methodology for environmental profiles of construction materials, components and buildings. BR370, 1999* A UK Ecopoint is a single unit measurement of environmental impact. An Ecopoint score is a measure of the total environmental impact of a particular product or process expressed in units (ecopoints). It is calculated in relation to impacts on the environment in the UK and therefore applies to UK activities only. Ecopoints describe all the environmental impacts arising from a product throughout its life cycle. They capture the relative importance which industry and society assigns to those environmental impacts.	✓	✓		✓		✓	✓				✓			✓	
EcoProp *http://research.scpm.salford.ac.uk* EcoProP is a requirements management tool, consisting of a generic classification of building properties, reference data about environmental requirements, information on relevant verification methods and automated procedures to scan requirements profiles and to form a design brief.	✓	✓			✓	✓					✓			✓	

Tool	Who will use this tool?				When will it be used?						Project types?		Relevance to WLV?		
	Designer	Client	Contractor	Other	Inception	Feasibility appraisals	Plan/design	Construct/handover	Operate/maintain	Decommission/renew	Buildings	Infrastructure	WLC	LCA	MCA
EcoQuantum *www.ivam.uva.nl/uk* Two versions of this Dutch LCA tool are available. Eco-Quantum Research is a tool for analysing and developing innovative and complex designs for sustainable buildings and offices. Eco-Quantum Domestic is a tool for architects to show environmental consequences of material and energy use of domestic building designs.	✓						✓				✓			✓	
EnergyStar / EPAStar *www.energystar.gov* This tool has been developed by the US Environmental Protection Agency. A range of on-line tools is available, and products (PCs, domestic appliances, and buildings) can be certified and awarded the EnergyStar logo if they meet the requirements of the scheme.	✓	✓		✓		✓	✓				✓			✓	
Engage: How to deliver socially responsible construction – a client's guide *www.ciria.org/acatalog/C627.html* Social responsibility is a key facet of sustainable development. It is concerned with addressing the needs of customers and shareholders, and all affected by an organisation's activities. This book provides guidance on building in a socially responsible way – putting inclusiveness, transparency and responsiveness at the heart of a construction project – while retaining a focus on business benefits. ENGAGE has two complementary parts: this guide and a web-based navigator.	✓	✓				✓	✓				✓	✓			✓
Envest *http://envestv2.bre.co.uk* A software tool to assess the life-cycle environmental impact of a proposed building and to explore design options and the impact of materials for the whole building. Also assesses the operational energy use and is particularly helpful at the early design stage.	✓					✓	✓				✓		✓	✓	
Environmental impact of building and construction materials *CIRIA, SP116, 1995. www.ciria.org.uk* Implications and uncertainties of eco-labelling and life-cycle assessment of building materials in general, followed by life cycle, energy and specification guidance for mineral products; metals; plastics and elastomers; timber and timber products; paints and coatings, adhesives and sealants.	✓					✓	✓				✓			✓	
Environmental Appraisal of Development Plans *www.nottingham.ac.uk/sbe/planbiblios/bibs/strategic/05.html* An explicit, systematic and iterative review of development plan policies and proposals to evaluate their individual and combined impacts on the environment.		✓		✓	✓	✓					✓	✓		✓	
Environmental Impact Assessment (EIA) *www.iaia.org/eialist.html* Initiatives for large-scale facilities and structures likely to cause significant environmental impacts are subject to the Environmental Impact Assessment Procedure in the EU (and quite widely worldwide).	✓	✓	✓		✓	✓	✓				✓	✓		✓	

Tool	Who will use this tool?				When will it be used?						Project types?		Relevance to WLV?		
	Designer	Client	Contractor	Other	Inception	Feasibility appraisals	Plan/design	Construct/handover	Operate/maintain	Decommission/renew	Buildings	Infrastructure	WLC	LCA	MCA
Framework for WLC *Email khu@gcal.ac.uk* Framework developed by John Kelly and Kristy Hunter of Glasgow Caledonian University on behalf of SCQS. It includes a CD with guidance manual and supported training face to face or web based.	✓			✓		✓	✓				✓	✓	✓		
Gateways Ahead *www.4ps.gov.uk/Documents/Publications/gateways_article_PFIIB.pdf* A Gateway Review is not an audit; it is a review of a project at key stages in its lifecycle, by independent 'peers' focussing on its likelihood to succeed. The process is designed to avoid any delay to a project with the confidential report being provided within the three-day review period. The Gateway service is now available to LAs through 4ps.		✓			✓	✓					✓	✓	✓		✓
GBTOOL *http://greenbuilding.ca/gbc98cnf/sponsors/gbtool.htm* A software tool that is under development as part of the Green Building Challenge process, an international effort to establish a common language for describing 'green buildings'.	✓	✓				✓	✓				✓			✓	
Good Practice Guides *www.energy-efficiency.gov.uk/index.cfm* These guides highlight the key points of a wide range of energy efficient technologies. They explain how they work, indicate how cost effective they can be and provide technical information to support users in preparing a case for funding such measures, and practical tips for implementation.	✓	✓			✓	✓					✓		✓	✓	
Green Guide to Specification (and for Housing) *http://products.bre.co.uk/breeam/greenguide.html* Simple ABC rating for specifiers selecting construction products on the basis of environmental impact over time.	✓						✓				✓			✓	
Impact matrix techniques *www.networkearth.org/naturalbuilding/matrix.html* Impact matrices are used to summarise the impacts of development activities in tabular form. Thus, columns in the table represent the alternatives of the project, plan, programme etc., and the rows represent the impacts considered. A matrix may contain an examination of the fulfilment of objectives (sustainability/quality of life), the need for mitigation measures or guidelines for subsequent planning phases, monitoring or follow-up needs, description of how the assessment has been carried out, assessment procedures required by legislation etc.	✓	✓	✓	✓	✓	✓	✓				✓	✓			✓
Impact matrix techniques *www.networkearth.org/naturalbuilding/matrix.html* Impact matrices are used to summarise the impacts of development activities in tabular form. Thus, columns in the table represent the alternatives of the project, plan, programme etc., and the rows represent the impacts considered. A matrix may contain an examination of the fulfilment of objectives (sustainability/quality of life), the need for mitigation measures or guidelines for subsequent planning phases, monitoring or follow-up needs, description of how the assessment has been carried out, assessment procedures required by legislation etc.	✓	✓	✓	✓	✓	✓	✓				✓	✓			✓

Tool	Who will use this tool?				When will it be used?						Project types?		Relevance to WLV?		
	Designer	Client	Contractor	Other	Inception	Feasibility appraisals	Plan/design	Construct/handover	Operate/maintain	Decommission/renew	Buildings	Infrastructure	WLC	LCA	MCA
KPIs for Customer satisfaction with the construction process and product *www.dti.gov.uk/construction/kpi* Construction Industry KPIs relating to performance in 2003 were launched in June 2004. The pack includes: handbook on the use of KPIs, a wallchart showing performance across the industry for 10 headline economic KPIs and a booklet on methods of measurement. Wallcharts and handbooks on Respect for People KPI and Environmental KPI, Specialist KPI wallcharts for consultants, M&E contractors and construction products. Additional Performance Indicators Book. Industry progress report. Case studies. Additional products and services Information Sheets. CD Rom.		✓		✓		✓	✓	✓	✓		✓	✓			✓
Life Expectancy of Building Components *www.bcis.co.uk/order/bmipub.html* RICS/ BMI questionnaire to surveyors about their expectations of the life expectancy of components, giving minimum, maximum and typical ranges of values for a wide range of components.	✓	✓	✓	✓	✓	✓	✓	✓	✓	✓	✓		✓	✓	
Local Government supplement to new SoPC 3 Guidance *www.4ps.gov.uk/Documents/Publications/SoPC%203%20Guidance.htm* Guidance on standardisation of PFI Contracts aimed at local authorities. Covers PFI, strategic service delivery partnerships, NHS LIFT and Building Schools for the Future programme briefly as well.		✓	✓	✓		✓	✓	✓	✓	✓	✓	✓	✓		✓
M4I EPIs (Environmental Performance Indicators) *www.constructingexcellence.org.uk/resourcecentre/kpizone* The following six indicators have been proposed by M4I: Operational carbon dioxide, embodied carbon dioxide, water, waste in the construction process, biodiversity, and transport.	✓	✓	✓				✓	✓		✓	✓			✓	
M4I Sustainable Construction Indicator *www.constructingexcellence.org.uk/resourcecentre/kpizone* Indicators have been developed for: area of habitat created/retained, commercial vehicle movements, energy use (construction process and product) impact on biodiversity and environment, water use (process and product) waste (process), whole life performance (product). They allow comparisons of project performance for various building types (offices, homes, hospitals, educational buildings, retail premises) against benchmarks and set targets for improvement.	✓	✓	✓				✓	✓	✓	✓	✓	✓		✓	✓
MaSC. Managing Sustainable Construction *http://projects.bre.co.uk/masc/index.html* MaSC is a management process underpinned by accelerated learning and a matrix for businesses to use to analyse their sustainability performance and set targets for continuous improvement. MaSC consists of two publications, MaSC: Profiting from Sustainability, an introduction to the process, and MaSC: Accelerated Learning, which guides companies through a series of structured actions to help manage the introduction of more sustainable practices. In house consultancy is also available to guide businesses through the process.	✓	✓	✓	✓					✓		✓	✓			✓

Tool	Who will use this tool?				When will it be used?						Project types?		Relevance to WLV?		
	Designer	Client	Contractor	Other	Inception	Feasibility appraisals	Plan/design	Construct/handover	Operate/maintain	Decommission/renew	Buildings	Infrastructure	WLC	LCA	MCA
MASTER (Managing Speeds of Traffic on European Roads) *www.its.leeds.ac.uk/facilities/lads/projects/master.html* The MASTER Framework is a set of guiding rules and principles for evaluating the impacts of a speed management policy so that the socio-economic feasibility of the policy can be established.	✓	✓		✓			✓		✓			✓			✓
NEAT *http://products.bre.co.uk/breeam/health.html* NEAT is checklist-based approach for assessing any new developments or refurbishment projects for NHS buildings. It aims to raise environmental awareness within the NHS and estimate the environmental impact and sustainability of NHS facilities and services. It scores on a scale from fail, to excellent.	✓	✓			✓	✓	✓			✓	✓			✓	
NHS LIFT and Local Government – Understanding Options: Considering Opportunities *www.4ps.gov.uk/Home.aspx?PageID=4.3.7&ctl=PubsDetails&PubsID=75* NHS LIFT is a Department of Health-sponsored initiative to develop and supply new and refurbished primary care facilities. It aims to increase awareness among local authorities of the NHS Local Improvement Finance Trust initiative to deliver joint health and local government activities, encourage local authorities to consider options for participating in NHS LIFT, explain its benefits and complexities; and outline its features compared to other forms of procurement.		✓			✓	✓					✓		✓		✓
Office Scorer *www.officescorer.info* The tool compares major or complete refurbishment with complete redevelopment. It enables users to systematically compare and test the environmental and economic impact of different building design concepts for offices and to identify sources of further relevant guidance. Its outputs are based on results that are derived from Envest, but in order to simplify the tool for users, a number of assumptions are made by the tool.	✓	✓			✓	✓				✓	✓		✓	✓	
Office, Schools and Local Authority Toolkits *EMAS or ISO 14001* The Toolkits are designed to help facilities, building or office managers to improve the environmental performance of their buildings, and indicate where these activities will help them save money. It is possible to use the toolkits in their own right or as a stepping stone towards a formal accreditation system, such as EMAS or ISO 14001.		✓		✓					✓		✓		✓	✓	
Outline Business Case, May 2004 *www.4ps.gov.uk/Documents/Publications/4ps%20Guidance%20-%20OBC%20-%2026.5.2004.pdf* This Note provides a framework for local authorities considering the procurement of a service or project through a PPP, strategic service partnership, PFI, or DBFO route. It promotes a structured and systematic approach to the development of the service or project and the Outline Business Case, and seeks to ensure that the decision to procure the project or service through a PPP or PFI route is based upon a robust strategic and financial analysis of the options available.		✓			✓	✓									

Tool	Who will use this tool?				When will it be used?						Project types?		Relevance to WLV?		
	Designer	Client	Contractor	Other	Inception	Feasibility appraisals	Plan/design	Construct/handover	Operate/maintain	Decommission/renew	Buildings	Infrastructure	WLC	LCA	MCA
Performance measurement for construction profitability *CT Cain, Blackwell Publishing. ISBN 1405114622. Available via www.brebookshop.com* Through performance measurement firms can identify their performance and unnecessary costs in the supply chain, leading to higher profits. The construction industry could meet the performance improvements demanded by end users, and replicate the gains of other sectors, using formal performance measurement. This guide focuses on the needs of managers at all levels. Using everyday business language, it explains how to set up and run performance measurement, self-assessment and benchmarking systems. It includes real-life examples.	✓	✓	✓	✓	✓	✓	✓		✓		✓	✓			✓
PFI in Schools – Quality and cost of buildings and services provided by Private Finance Initiative schemes *www.audit-commission.gov.uk* PFI is increasingly important in the provision of public sector facilities. To date, only a small number of new PFI schools have been completed, but commitments have been made to build many more. It is crucial that these schools get the most from PFI, and that early lessons are recycled effectively during future investment. This report gives a detailed overview of all aspects of PFI in schools.	✓	✓	✓		✓	✓	✓				✓		✓		✓
PPP COM – Integrated risk and value management toolkit for Public Private Partnerships *I Cruickshank et al. (2003). CIRIA C617 www.ciria.org/acatalog/C617CD.html* The toolkit offers a common platform and a common language for the communication of risk issues. It helps users to improve the quality of risk and value management on their projects by providing an on-the-job educational tool for risk management. PPPCom is a readily transferable process, as 80% of the information is generic and so may be exported to other industries, with 20% aimed specifically at construction projects.	✓	✓	✓	✓	✓	✓	✓	✓	✓	✓	✓	✓	✓		✓
Reputation, Risk and Reward – report by Sustainable Construction Task Force, BRE *http://projects.bre.co.uk/rrr/index.html* This paper is addressed primarily to chief executives, board members and senior managers in property development and property asset management; as their business partners and key suppliers, it should also be of interest to executives in the construction sector. The purpose of the paper is to demonstrate that sustainability is an issue that business leaders can no longer afford to ignore or treat as immaterial. This paper aims to demonstrate that sustainability issues are of critical and strategic importance to business.		✓		✓	✓	✓			✓	✓	✓	✓		✓	✓
RESUS (Recycling and sustainability in civil engineering) *www.resus-engineer.co.uk* RESUS is a research and development project on recycling and sustainability in civil engineering, being undertaken by Waterman Burrow Crocker and supported by ICE. It includes a generic Decision Support Tool for development of specifications and standards for the use of sustainable materials in civil engineering. The generic decision processes have been translated into examples for highway pavement and drainage designs, and are available from the project website. Whole life value aspects are integrated into the processes under the heading performance requirements.	✓	✓	✓	✓	✓	✓	✓	✓	✓	✓	✓	✓	✓	✓	✓

Tool	Who will use this tool?				When will it be used?						Project types?		Relevance to WLV?		
	Designer	Client	Contractor	Other	Inception	Feasibility appraisals	Plan/design	Construct/handover	Operate/maintain	Decommission/renew	Buildings	Infrastructure	WLC	LCA	MCA
SIGMA project (Sustainability - Integrated Guidelines for Management) *www.projectsigma.com* An integrated system of guidelines for management of sustainability issues within organisations irrespective of size or sector. The SIGMA project, launched in 1999 with support from the Department of Trade and Industry, is a partnership between the BSI (the standards organisation), Forum for the Future (a sustainability charity and think-tank), and AccountAbility (the international professional body for accountability). Training is available to implement the guidelines. www.bsi-global.com/Seminars/SIGMA/index.xalter has more detail.	✓	✓	✓	✓	✓	✓					✓	✓			✓
SMARTWaste & SMARTStart, Waste auditing software, BRE *www.smartwaste.co.uk/smartstart/aboutsmartstart.jsp* The true cost of construction waste on-site is not only the cost of the skip. It includes wasted resources and time and the cost of disposal. Reducing waste can significantly improve profits and reduce environmental impacts. SMARTStart supplies an easy-to-use benchmarking tool for a whole company –an ideal first step on the road to waste reduction on site. SMARTWaste will empower a company to understand in detail the waste generating processes on site and enable to implement and monitor waste management and minimisation plans.			✓					✓		✓	✓	✓	✓	✓	
Social Cost-benefit Analysis (SCBA) *http://research.scpm.salford.ac.uk* SCBA identifyies the actions that either minimise the social costs when outputs (objectives) are given or maximise the output (achievement of objectives) within a given budget; and establishing the social distribution of the impacts.		✓		✓	✓	✓	✓			✓	✓	✓			✓
SPARTACUS (System for Planning and Research in Towns and Cities for Urban Sustainability) *http://research.scpm.salford.ac.uk/bqtoolkit/tkpages/ass_meth/methods/amspa_4.html* SPARTACUS is an indicator system and a decision support tool for assessing sustainability implications of urban land use and transport policies. It is based on the results of a transport land use interaction model.		✓		✓	✓	✓	✓					✓			✓
SpEAR (Sustainable Project Appraisal Routine) *www.arup.com/sustainability/services/SPeAR.cfm* A project appraisal methodology, to be used as a tool for rapid review of the sustainability of projects, plans, products and organisations. SpEAR allows the sustainability of a project to be assessed and illustrated graphically at all project stages, demonstrating continual improvement and evolution of a project over time. It allows the various aspects of sustainability to be balanced and their inter-relationship assessed.	✓	✓	✓		✓	✓	✓				✓	✓		✓	✓
Spreadsheet modelling in investments – A book and CD series *Craig W Holden. Prentice Hall 2002 ISBN 0-13-087948-7* This comes a book and browser-accessed CD-ROM that teaches students how to build financial models in Excel. Provides instructions for building financial models, not templates. Progresses from simple examples to complex real world applications.				✓	✓	✓	✓				✓	✓	✓		

	Who will use this tool?				When will it be used?						Project types?		Relevance to WLV?		
Tool	Designer	Client	Contractor	Other	Inception	Feasibility appraisals	Plan/design	Construct/handover	Operate/maintain	Decommission/renew	Buildings	Infrastructure	WLC	LCA	MCA
Sustainability Accounting *www.forumforthefuture.org.uk/SustainabilityAccounting_page1352.aspx* and *www.projectsigma.com/Toolkit/SustainabilityAccountingGuide.asp* Financial accounting traditionally records the financially related flows and stocks of an organisation as profit and loss account and balance sheet. The key stages in developing external cost accounts under the Forum for the Future methodology are: identification and confirmation of the organisation's main environmental impacts; estimation of what a sustainable level of impacts may be to determine sustainability targets or the 'sustainability gap'; valuation of those impacts – on the basis of what it would cost to avoid them in the first place, or if avoidance were not possible, what it would cost to restore any resulting damage (using market-based prices where possible). The tool is called SIGMA.	✓	✓		✓	✓	✓	✓				✓	✓		✓	✓
Sustainability Checklist for Developments, BRE *BRE Report 436. www.brebookshop.com* One of the biggest challenges facing developers, designers and planners is ensuring that our towns and cities are developed and regenerated to be sustainable for the future. This Checklist provides practical tools and indicators to measure the sustainability of developments (both buildings and infrastructure) at site or estate level, and a common framework for discussions between developers, local authorities and communities.	✓	✓	✓	✓	✓	✓					✓			✓	✓
Sustainability lessons from PFI and similar private initiatives *http://projects.bre.co.uk/sustainabilitylessons/publications.html* This report discusses case studies that are the output of a project to establish lessons from PFI and similar initiatives. The report identifies the need for a vision and commitment to sustainability from all parties. The whole team, from client to contractor must 'buy in'. It also claims that poor communication between the team will make sustainability difficult to achieve, it is important to get all disciplines involved. The report also provides a summary of the potential opportunities and pitfalls for sustainability for PFI/PPP projects.	✓	✓	✓	✓	✓	✓	✓	✓	✓	✓	✓	✓			✓
Sustainability Works - The complete development tool for sustainability housing *www.sustainabilityworks.org.uk/sus.php* Sustainability Works is a powerful online application that combines a comprehensive information resource with practical support for successful planning, implementation and monitoring of sustainable housing policies and projects. It is organised around the development process and covers eight key themes: buildings, energy, land use, landscape, society, travel, waste and water. Case studies, costs and benefits, links to information, and references to further research are included for recommended policies and measurable targets.	✓	✓	✓		✓	✓	✓	✓			✓			✓	✓

Tool	Who will use this tool?				When will it be used?						Project types?		Relevance to WLV?		
	Designer	Client	Contractor	Other	Inception	Feasibility appraisals	Plan/design	Construct/handover	Operate/maintain	Decommission/renew	Buildings	Infrastructure	WLC	LCA	MCA
Sustainable construction procurement *By B Addis and R Talbot. CIRIA. ISBN 0 86017 571 5. Available via www.brebookshop.com* This 'Guide to delivering environmentally responsible projects' will help those engaged in construction projects to improve their environmental and sustainability performance, for both the product and process of construction. It gives an overview of environmental responsibility and sustainability, and advice at all stages of projects. It summarises the experience of organisations that have tackled these questions, and identified key issues, legislation, guidance and pitfalls, and successful techniques for delivering projects encouraging environmental responsibility. The guide contains case studies, background information, advice on key actions and routes to sources of information and guidance.	✓	✓		✓		✓	✓	✓	✓	✓	✓			✓	
The Green Guide to Specification *BRE Report 351, 1998. www.brebookshop.com* An Environmental Profiling System for Building Materials and Components. The Green Guide provides a pointer to specification, which is both easy to use and soundly based on numerical data. These measure the environmental impacts of building materials in terms of 13 key parameters based around embodied energy, emissions, toxicity, wastes, and use of resources	✓					✓	✓				✓			✓	
The Natural Step *www.naturalstep.org* The Natural Step engages with companies to transform the way they do business by integrating sustainability principles into their core strategies, decisions, operations and bottom line. We help companies understand how the current state of the world is impacting business systems, and how, in turn, business systems are impacting the state of the world.	✓	✓	✓	✓	✓	✓	✓	✓	✓	✓	✓	✓		✓	
Twenty steps to encourage the use of Whole Life Costing – rethinking construction *www.constructingexcellence.org.uk/sectors/housingforum/document.jsp?documentID=115540* The Whole Life Costing working group of the Local Government Task Force has drawn together a wide range of experienced practitioners and academics along with IT system suppliers, to prepare this report. The group has consulted local authorities, housing associations, house builders, contractors, consultants and PFI organisations in drawing up the 20 tips. This report is a starting point in helping clients and others to take the first steps in adopting a structured management approach to efficient long-term investment in housing.	✓	✓	✓			✓	✓				✓		✓		
VALiD (VALue in Design) *www.valueindesign.com* VALiD is an approach to helping stakeholders understand the through-life value of a design solution. Supported by an IT tool to help stakeholders express their judgements of design performance, VALiD is a continuous process that considers value to arise from the relationship between the benefits and sacrifices associated with a construction projects and product use. VALiD allows value delivery performance to be measured by comparing stakeholders' judgements of design performance against those set at the project outset.	✓	✓	✓	✓	✓	✓	✓	✓	✓	✓	✓	✓			✓

	Who will use this tool?				When will it be used?						Project types?		Relevance to WLV?		
Tool	Designer	Client	Contractor	Other	Inception	Feasibility appraisals	Plan/design	Construct/handover	Operate/maintain	Decommission/renew	Buildings	Infrastructure	WLC	LCA	MCA
Waste minimisation in construction, training pack *www.ciria.org.uk* This pack for those training construction staff includes: Users instructions, trainer's notes, quiz, a CD containing: Powerpoint presentation - can be printed for trainees, group exercise. C536 Demonstrating waste minimisation benefits in construction (includes 10 case studies); SP148 Waste minimisation in construction 15-min video; SP113 Waste minimisation in construction site guide. It is relevant to all levels of staff. It is flexible and can be adapted for training sessions of different lengths, and for audiences of different types.			✓					✓			✓	✓		✓	
Waste minimisation in construction – site guide *www.ciria.org.uk* This comprehensive guide on waste minimisation measures is aimed at policy-makers of all organisations in the various roles in the construction industry including clients, designers, contractors and suppliers.	✓		✓				✓			✓	✓	✓		✓	
Whole Life Cost of Social Housing *www.housingcorplibrary.org.uk/housingcorp.nsf/AllDocuments/52474CEAE6829ADE80256DCC00530156* BCIS study on whole life costs of social housing commissioned by the Housing Corporation, October 2003.	✓	✓	✓	✓	✓	✓	✓	✓	✓	✓	✓		✓		
Whole life costing approach to water distribution network management *P Skipworth, A Cashman et al. Thomas Telford. Available via www.brebookshop.com* This book presents a solution to problems of efficiently investing in deteriorating water distribution networks. When applied to service infrastructure, WLC can be used to consider all the costs of initiation, provision, operation, maintenance, servicing and decommissioning, over its useful life. Detailed within this book is a WLC methodology for water distribution networks. These drivers are often a function of network performance and the service received by the customer. Thus, the book is a reference for both the cost accounting and performance modelling of water networks. It demonstrates the application of the WLC methodology within the regulated water industry in England and Wales.		✓	✓			✓	✓	✓	✓	✓		✓	✓		✓
Whole Life Value Framework – web-based search engine *www.wlv.org.uk* Search tool described in BRE IP 10/04, designed to allow users to identify relevant sources of information, tools and guides across a range of criteria and building stages.	✓	✓	✓	✓	✓	✓	✓	✓	✓	✓	✓	✓	✓	✓	✓
Whole life-cycle costing: risk and risk responses *R Kirkham, H Boussabaine. Blackwell, 2004. www.brebookshop.com/details.jsp?id=143660*	✓	✓	✓	✓	✓	✓	✓	✓	✓		✓	✓	✓	✓	

References

Association for Project Management (1997) Project Risk Analysis and Management (PRAM) guide, available from www.apm.org.uk.

British Airports Authority (1996) 21st Century Airports – aiming for world class standards in capital investment, BAA plc, 130 Wilton Road, London, SW1 1LQ.

CPN (2004) Local Government procurement: Delivering best value in practice, Construction Productivity Network (CPN) Workshop Report, E4126, available from www.ciria.org.

Confederation of Construction Clients (2000) Clients' Charter, available from www.clientsuccess.org.uk.

Construction Clients' Forum (1999) Whole Life Costing – a client's guide, available from www.clientsuccess.org.uk.

Connaughton J. and Green S. (1996) Value management in construction: a client's guide, CIRIA SP129, available from www.ciria.org.

Cruickshank I et al. (2003) PPP COM – the integrated risk and value management toolkit for Public Private Partnerships, CIRIA C617, available from www.ciria.org.

Department for Transport (2000) Guidance on the Methodology for Multi-Modal Studies, available from www.dft.gov.uk.

DETR (1998) A New Deal for Trunk Roads in England: Understanding the New Approach to Appraisal, HMSO, London.

DETR (2000) Building a Better Quality of Life: A Strategy for more Sustainable Construction, HMSO, London.

DETR. MCA Manual. http://www.odpm.gov.uk/stellent/groups/odpm_about/documents/page/odpm_about_608524.hcsp

Egan Sir J. (1998) Rethinking Construction, Construction Task Force Report, HMSO, London.

Gershon P. (1999) Review of Civil Procurement in Central Government, Office of Government Commerce, London.

Godfrey P.S. (1996) Control of risk: a guide to the systematic management of risk from construction, CIRIA Special publication 125, CIRIA, available from www.ciria.org.

Government Construction Clients' Panel (GCCP) (2000) Achieving sustainability in Construction Procurement, GCCP, London, available from www.ogc.gov.uk.

Highways Agency (2004) Stonehenge WHS Management Plan, available from: www.highways.gov.uk/roads/projects/a_roads/a303/stonehenge/index.htm.

HM Treasury (2003) Appraisal & Evaluation in Central Government: the Green Book, The Stationery Office, London.

HM Treasury (2000) Procurement Guidance Note No. 7 on Whole Life Costs. November, available from www.ogc.gov.uk.

Holti, R et al. (2000) The handbook of supply chain management, CIRIA C546, available from www.ciria.org.

Institution of Highways and Transportation (2001) Delivering Best Value in Highway Maintenance: Code of Practice for Maintenance Management, DETR, London.

International Financial Services (2003) PFI in the UK: progress and performance, PPP Brief, London.

International Standards Organisation (2001) Buildings and Constructed Assets – Service Life Planning: Part 1 General Principles, Final International Standard. ISO/DIS 15686-1, available from www.iso-14001.org.uk.

International Standards Organisation ISO 14040 Life Cycle Assessment, Guidance on the ISO 14000/14001, series of standards, available from www.iso-14001.org.uk.

Latham, M. (1994) Constructing the Team, HMSO, London.

National Audit Office (2001) Modernising Construction, available from www.nao.gov.uk.

National Audit Office and Office of Government Commerce (2001) Getting Value for Money from Procurement: How auditors can help, Available from www.nao.gov.uk and www.ogc.gov.uk.

ODPM (2004) Multi-criteria assessment – a manual, Office of the Deputy Prime Minister. www.odpm.gov.uk/stellent/groups/odpm_about/documents/page/odpm_about_608524.hcsp.

Office of Government Commerce (1999) Achieving excellence, available from www.ogc.gov.uk.

Society of Environmental Toxicology and Chemistry (SETAC), 1999.

Strategic Forum for Construction (2002) Accelerating Change, available from www.constructingexcellence.org.uk.